ADVANCE PRAISE FOR

## *What We Think About When We Try N*
## *To Think About Global Warming*

"The human brain is poorly equipped to cope with mind-numbing problems like climate change. Per Espen Stoknes tell us why—and then explains what we can do to change the way we think, act, and live. Highly recommended."
— **John Elkington**, cofounder of Volans, SustainAbility, and Environmental Data Services (ENDS) and coauthor of *The Breakthrough Challenge*

"How, most effectively, to communicate the reality and ramifications of a slow-motion planetary meltdown? Whether you are a scientist or a CEO, an activist or a slacker, this book provides a simple toolkit for breaking down frozen attitudes. As a work that surveys a great deal of psychological research, it's at once accessible, practical, and—in its last third—richly reflective and evocative. In these concluding chapters, Stoknes wrestles eloquently with the ways in which earthly calamity reverberates and sometimes wreaks havoc in any person's innermost sense of self and meaning."
— **David Abram**, author of *The Spell of the Sensuous*

"In a fresh and intimate voice, Per Espen Stoknes navigates the obstacles and collective denial of climate change. Drawing on his own deep love of nature, he suggests ways to overcome our 'Deep Grief' by creating a spiritual connection with the air around us. In every way, this is a book full of new perspectives and insights."
— **George Marshall**, author of *Don't Even Think About It: Why Our Brains Are Wired to Ignore Climate Change*

"If information enlightened, then effective climate policies would have been put in place two decades ago, after the second IPCC assessment. The recent, massive fifth assessment enlightens only a teeny bit more. Stoknes's small, powerful, readable book enables us to build the social networks that will lead to action and change our old stories, the blinders that comfort so many along our path to destruction. Read it, get to work, and find joy in being effective."
— **Richard B. Norgaard**, coauthor of *The Climate Challenge Society* and professor emeritus, University of California at Berkeley

"Mahatma Gandhi said 'First they ignore you, then they laugh at you, then they fight you, then you win.' We're in this last phase, but to win we need to change tactics, from using guilt to draw attention, to instead using persuasion to change behavior and policy at a mass scale. Per Espen Stoknes shows the way with this brilliant description of how to go *with* rather than *against* the flow of human nature and thus shift society to action. There is no more important challenge facing society today and Stoknes's contribution is crucial."

—**Paul Gilding**, author of *The Great Disruption*

"Science is no longer the bottleneck to action on climate change. Why do we so often ignore, deny, and resist the science? Why aren't we outraged, demanding change? In a style both rigorous and personal, Per Espen Stoknes explains why, and, more importantly, offers strategies for success. A pleasure to read, this book can help us all become more understanding, more committed, more effective—and, along the way, more joyful."

—**John Sterman**, MIT Sloan School of Management and author of *Business Dynamics: Systems Thinking and Modeling for a Complex World*

"Stoknes offers expert insights, drawn from the discipline of psychology and the art of storytelling, to the high-stakes quandary of our time: why the response to climate change has not, yet, come close to matching the overwhelming magnitude and sophistication of the scientific evidence. He peels away the multiple layers of passivity-inducing narratives and demonstrates how avoiding climate carica-tures—apocalypse on one hand, ecotopia on the other—is the most effective way to prompt action. His alternative narratives, highlighting the many co-benefits of a switch away from fossil fuels, suggest a broad common ground across the ideological spectrum."

—**Mark Schapiro**, author of *Carbon Shock: A Tale of Risk and Calculus on the Front Lines of a Disrupted Global Economy*

"Combining an entrepreneur's innovation with an economist's analytics and a psychologist's knowledge of human behavior, Per Espen Stoknes gives us a much-needed guide to moving beyond the politics and paralysis that generally cripple action on climate change and provides us with concrete ways to inspire grounded hope for real climate solutions."

—**Heidi Cullen**, chief scientist, Climate Central

# What We
# Think About
### When We Try Not To Think About
# Global
# Warming

# What We Think About

## When We Try Not To Think About

# Global Warming

*Toward a* **NEW PSYCHOLOGY** *of Climate Action*

## Per Espen Stoknes

*Foreword by* **Jorgen Randers**

Chelsea Green Publishing
White River Junction, Vermont

Editor: Joni Praded
Project Manager: Bill Bokermann
Copy Editor: Laura Jorstad
Proofreader: Helen Walden
Indexer: Peggy Holloway

Printed in the United States of America.
First printing March, 2015.
10 9 8 7 6 5 4 3 2 1     15 16 17 18

**Our Commitment to Green Publishing**
Chelsea Green sees publishing as a tool for cultural change and ecological stewardship. We strive to align our book manufacturing practices with our editorial mission and to reduce the impact of our business enterprise in the environment. We print our books and catalogs on chlorine-free recycled paper, using vegetable-based inks whenever possible. This book may cost slightly more because it was printed on paper that contains recycled fiber, and we hope you'll agree that it's worth it. Chelsea Green is a member of the Green Press Initiative (www.greenpressinitiative.org), a nonprofit coalition of publishers, manufacturers, and authors working to protect the world's endangered forests and conserve natural resources. *What We Think About When We Try Not To Think About Global Warming* was printed on paper supplied by McNaughton & Gunn that contains 100% postconsumer recycled fiber.

**Library of Congress Cataloging-in-Publication Data**
Stoknes, Per Espen.
What we think about when we try not to think about global warming :
toward a new psychology of climate action / Per Espen Stoknes.
     pages cm
ISBN 978-1-60358-583-5 (paperback) -- ISBN 978-1-60358-584-2 (ebook)
1. Climatic changes--Psychological aspects. 2. Global warming. 3.
Environmental policy. 4. Environmental psychology. I. Title.

BF353.5.C55S76 2015
155.9'15--dc23

2014047859

Chelsea Green Publishing
85 North Main Street, Suite 120
White River Junction, VT 05001
(802) 295-6300
www.chelseagreen.com

To Sølve and Samuel,
to all who feel the unsettling winds of change,
and to those who are yet to be born into the air

# —CONTENTS—

Foreword      ix

Acknowledgments      xi

Introduction: Battering One Another      xiii

## Part I. Thinking
### *Understanding the Climate Paradox*

1. The Psychological Climate Paradox      3
2. "Climate Is the New Marx": The Many Faces of Skepticism and Denial      9
3. The Human Animal, as Seen by Evolutionary Psychology      27
4. How Climate Facts and Risks Are Perceived: Cognitive Psychology      35
5. What Others Are Saying: Social Psychology      54
6. The Roots of Denial: The Psychology of Identity      70
7. The Five Psychological Barriers to Climate Action      81

## Part II. Doing
### *If It Doesn't Work, Do Something Else*

8. From Barriers to Solutions      87
9. The Power of Social Networks      95
10. Reframing the Climate Messages      110
11. Make It Simple to Choose Right      124
12. Use the Power of Stories to Re-Story Climate      132
13. New Signals of Progress      151

## Part III. Being
### *Inside the Living Air*

14. The Air's Way of Being      165
15. Stand Up for Your Depression!      171
16. Climate Disruption as Symptom: What Is It Trying to Tell Us?      190
17. Re-Imagining Climate as the Living Air      202
18. It's Hopeless and I'll Give It My All      217

Notes      229

Bibliography      251

Index      277

Forty years ago, I worked on the MIT team that created *The Limits to Growth*, the report that caused a global debate that persists today. We had projected possible futures for a world that would be living beyond what its resources could sustain, and the message was not an easy one: endless physical growth was not possible, in the long term, on a finite planet. Our model showed that eventually, when pushed to its limit, the system that supports life on earth would crash. Unless we took steps toward sustainability.

Ever since, I have been working to make sustainable development a reality, and as I see it, with very little success. The world is less sustainable today than it was in 1970 when I started my endless effort. We seem unable to act on what we know about realities like climate change, shrinking biodiversity, and depleting mineral resources. In 2011, I finally gave up, and wrote the book *2052: A Global Forecast for the Next Forty Years*. In it, I describe what, given what I now know about human nature, will likely happen toward the middle of this century if global society continues not to listen to science-based advice. It is not an attractive future, and it makes me depressed to be reminded every day that I was not able to convince the world to create a much better one. Especially since this is fully possible, with limited sacrifice.

I share this so that you can understand why I am simply delighted that I agreed to write a foreword for Per Espen Stoknes's book on climate psychology. This forced me to read the book first, and to my great surprise and pleasure, the reading changed my life. Or, to be more precise: it changed my mental frame from depression to hope.

Per Espen's message was not new to me. I have been a close colleague of his for fifteen years or so. But reading his new book did something to me that he has not been able to do orally over all these years. Reading his book gave me back the hope I had gradually lost during my forty years of futile struggle.

I now understand why my "rational" strategy during all those years failed, and is bound to keep failing. Pushing ever more information onto people who are in denial will never help. It will never help to further clarify the climate damage we are likely to face, and the sacrifice (albeit small) that would be needed to solve the problem. Instead, we must make it more attractive in the short term for a large group of citizens to take part in the creation of a new green future than to remain in the old fossil one.

This can be done, says Per Espen, and I now agree, by arranging a new reality, where there is an opportunity for you to reduce your energy use in friendly competition with your neighbors. Where suppliers systematically make the green option the default choice. And where I, in my public talks, change from the apocalypse story to the happy story of economic revitalization, happiness, social justice, a good life, stewardship, and re-wilding. The happy version of a possible future is as real as the apocalyptic version—but much more inspiring.

For realists like myself, Part III of the book is probably the most intriguing and helpful. I have felt for decades the burden of "Big Grief" when nature was increasingly destroyed around me. But I thought it was only me who felt that way. What a relief to hear Per Espen suggest the totally obvious fact that we humans are part of a totality (nature) and equally obviously are bothered (plagued and tormented, really) when parts of this totality are destroyed. What a pleasure to discover the hope that rests in the fact that this dissonance can be solved if we stop destroying nature. And, since stopping the destruction is the only way to remove the dissonance, it is likely to happen one day. We humans will remain in nature, and stop hurting it.

JORGEN RANDERS

# —ACKNOWLEDGMENTS—

My best moments come when I dwell in the house of gratitude. Appreciation and acknowledgments flow in so many directions, including to the air itself for the nourishment and beauty received through its light and restless flows. Clouds, pines, and sparrows have silently held the space around my writing cabin at Meheia, Norway, allowing words to flow through and settle down on the pages. Thanks, too, for the courage to cut and delete the superfluous fragments, which may then go and show up elsewhere. Writing may seem to be a solitary process, but—in other ways—it is an intimate immersion in a wider web of lives and intriguing ideas.

Thus, this book synthesizes research from a wide span of disciplines, stands on many years of collaboration with colleagues, and draws on the work of many hundreds of courageous climate scientists, social scientists, psychologists, journalists, and communication professionals, as witnessed by the literature list. Without their work, insights, and contributions, *What We Think About When We Try Not To Think About Global Warming* could not exist.

The idea for the book came out of my collaboration with Jorgen Randers, with whom I shared a climate-strategy professorship at the Norwegian Business School's Center for Climate Strategy, made possible by a grant from Toyota Motors. The insight about the need for the book came during one meeting in 2009, when we discussed the "decision-problem"—the reality that the lack of effective climate response is due not to a lack of facts or technology or policy solutions, but to "softer," psychological reasons. At the time, no book was available on this topic.

I'd like to send thanks to the many audiences that have given their enthusiastic feedback to my keynotes and seminars on climate psychology. Further, I'm really grateful to these wonderful folks for conversations, ideas, feedback, and help and assistance in writing, editing, and discussing the many drafts of the book, in no particular order: Scott Becker, David Abram, Stephan Harding, Per Ingvar Haukeland, Anthony Leiserowitz,

Elke Weber, Kari Marie Norgaard, Richard Norgaard, George Lakoff, Heidi Cullen, Heather Pittman, Roger Stephenson, John Grim, Mary Evelyn Tucker, Jerome Bernstein, Richard Tarnas, Elizabeth Allison, Maximilian DeArmon, Theo Badashi, Larissa Stendie, Steffen Kallbekken, Bodil Fisknes, Sigrid M. Hohle, Jarle Fagerheim, Kari Bu, Ole Christian Johnsen, Silje Wästlund, Stein Stoknes, Bjørnar Berg, Markus Lindholm, Vita Galdike, Anne Jortveit, Anders G. Imenes, Ingunn Grande, Otto Simonett, Daniel Erasmus, Ulf Myrvold, Liz Rose, Helene Høye, Nikyta Palmisani, and more that may have slipped out of this list.

Thanks to Tone W. Melbye for assistance in typing the manuscript.

Grateful appreciation to Joni Praded, my amazing senior editor, for her systematic eye and pen, helping out with my Norwegian-English idiosyncrasies, as well as to the whole team at Chelsea Green Publishers.

Thanks, too, to my sons, Sølve and Samuel, for being such amazing sons even through periods with a seriously absent-minded father.

Above all, thanks to my wife, Anne Solgaard, who has unwaveringly reminded me of prioritizing my calling, and kept up her support for me when writing this book through life's many ups and downs.

# Battering One Another

I was talking to a group of forty senior industry executives when the air in the hotel conference room began to feel charged. No more than ten minutes into my talk on climate psychology, I sensed a brewing discomfort in my stomach. Then one of their leading members cut me off from his first-row seat. "This global warming thing you're talking about is very uncertain," he declared. "It's been hyped. There is even a Nobel Prize winner in physics, Professor Ivar Giaever, that has documented that global warming isn't happening."

Such comments are not uncommon. I wasn't very surprised. This topic was hot, the audience was feeling criticized by the global warming message, and I was being challenged. The glove had been thrown down. What could I say that they would hear? To go on with my next slide was not an option. The challenge couldn't just be ignored. Those in the group were mostly white, mostly male, and mostly in suits and ties, like myself. The questioner's voice was friendly, but I could feel the hostility mounting just under the surface.

Something in the climate message is unsettling, maybe even *disarranging* to our minds. And in that hotel conference room, on a cool November day, the psychology of climate was being enacted in real time—ironically during my lecture on the very same subject.

No wonder there is a gut reaction to shoot down the message. Or the messenger. Take "them" down. Kill the doom-mongering. After all, argument is war, and the climate debate has been exactly that. Just think about the word *debate*. The prefix *de* refers to "down," as in *de*pression. And *bate*— as in *bat*, one with which to hit the opponent—comes from the Latin word *battere*, to fight. So debate is fighting by hitting someone down with a bat.

The climate debate has devolved into just such a deteriorating and desperate spiral. Many of those who doubt global warming are now seeing the record numbers of people hitting the streets demanding climate action. They may be noticing that large accounting firms, insurance associations, and military analysts are increasingly ranking climate as a serious risk to the economy and security. And they may even be experiencing firsthand heavier floods, longer droughts, or other effects of climate change. But even so, the debate rages on or is carefully avoided altogether.

How much longer will so many people feel a need to combat climate and ecological science? Climate change is now a more divisive topic in the United States than abortion, gun control, the death penalty, or genetically modified crops.[1]

If you want change to happen, hard confrontation is usually not productive. Coaches and psychotherapists know this. You don't debate your client in the hope that you—the coach—win and they lose. The more you try to force change onto someone, the tougher the resistance. Getting into us-versus-them positions invites more ditch digging, not dialogue. The stronger argument rarely wins in practice, unless the opponent really wants to learn and explore. And even if you win today, the losers live to fight another day. They don't go home and change their minds in wise reflection. I, at least, don't change that way.

Change can happen through dialogue, but what is needed first is curiosity, empathy, and focus on finding some common ground.

For those of us who feel the unsettling winds of change in the air—literally—the time for debating with contrarians about whether climate change is happening is over. It's no use trying to win the argument against those that have made up their minds to the contrary. We don't even want to win over the climate contrarians so that they lose. What we want is movement out of the trenches. Attention. Common ground. Joint exploration of solutions. A shift toward new stories. And for that to happen, we first need to understand the internal resistance, the deadlock: What's holding back the long-overdue shift? What's stopping the facts of climate change from unsettling our minds enough that the much-needed swerve in public opinion can happen?

So what did I reply to the group of executives? That I'm *happy* that up to 2 or 3 percent of climate researchers insist on thinking differently.[2] Living with diversity may be taxing, but the alternative is usually worse. Diversity of species and diversity of minds are both indispensable to a vibrant earth as we move deeper into the twenty-first century. The executives could accept this line of reasoning; they know that having sufficiently diverse mind-sets when doing strategic analysis is critical to avoid groupthink in executive teams. We could proceed to the issues of fear messaging, worldviews, and new opportunities of green growth.

But being happy about the thin, thin 2 to 3 percent slice of contrarian researchers does not give license to ignore the other 97 percent—even if ignoring the unsettling climate facts could be personally more convenient for my lifestyle. Choosing ignorance would let me off the hook of feeling implicated when I fly too much—or don't contribute enough.

The basics of greenhouse gas atmospheric warming are simple. The presence of certain gases traps more of the sunlight's heat close to earth, so that less is radiated back out into space. Since the industrial revolution, our species has been releasing more of these gases into the air than are taken out by natural processes. This shifts the earth's energy balance away from the delicate stability during which human civilization has flourished, leading to disruptive conditions for humans and today's ecosystems. That's about it. Three sentences.

But the devil is in the details. Climate sciences at the planetary scale quickly become extremely complex. Since the ever-moving air is linked to rainfall, to clouds, to ground, to ocean, to chemistry, physics, biology, ecology—everything really—the issue starts spanning many, many disciplines. A person really needs years and years of dedicated training in order to understand just one of these disciplines at some depth. And this is even before we include human societies and the social sciences. In democracies, the question becomes what people should *believe* about this complexity. Which experts are to be trusted?

What we see at this point is that around half of the population in rich countries chooses to side with the tiny sliver of 2 percent rather than the 98 for some reason or other.[3] That is a fascinating paradox. Pretty scary, actually. And therefore worth thinking deeply about. What parent can today take to heart that within the lifetime of their newborn baby, the planet will become hotter than it's been in millions of years?

This book is about such paradoxes, responses, and solutions. Understanding human responses to climate change is clearly becoming just as important as understanding climate change itself. The main question is: What do our reactions to the climate change facts tell us about the way we think, what we do, and how we live in the world? And how can we use what we know about human nature—our own, and others'—to move beyond our psychological barriers to making a great climate swerve?

## Thinking About the Future

Several years ago, not long after my divorce, I found myself walking back from my now ex's new apartment, having left our kids at her place. Walking away alone. Our life project of the previous fifteen years was in ruins. Now what to do with my life? The pain from my broken dreams was as palpable as the freezing-cold night around me. I noticed the winter stars, and the waning old moon lingering just above the horizon, to the left of a tall office building next to my new attic in central Oslo.

Then something unexpected happened. I could feel the air around me like never before. It seemed to descend from the sky itself, flow down from the moon, and rise up from the ground. It enveloped my hair, chilled each finger as I swung my arms and walked in its flow. I was walking with pain in my heart, but something new was opening before me, too, and I reached a decision I didn't even know I was contemplating. The rest of my life, I decided right then and there, would revolve around climate-related work.

If you're thinking you've just opened a book by a new-age evangelizer, you can relax. The climate epiphany didn't come out of nowhere. I had grown up in a family-owned, smelly fish factory on Norway's gorgeous western fjord coast and later exercised the entrepreneurial and curious genes spawned there in the worlds of green tech and plasma physics. I had become a certified psychologist with a PhD in economics, and had been exploring future strategic scenarios, consulting across four continents, and wondering how my worlds of science and storytelling, therapy and policy, climate reality and human imagination might eventually collide. But I hadn't yet accepted climate as a defining feature in my life's work.

It has taken some years to digest the answer that seemed to come right out of the open air that evening. Does the act of accepting personal loss open the heart to the more transpersonal pain of displaced peoples, forests and ocean, the furred and the finned ones? It certainly seemed that way. And it makes more sense each time and every day I reflect on it. Acknowledging distress on one level led me to connect better to it on another.

In the years since, much of my climate-related work has been in collaboration with my older colleague Jørgen Randers, who back in 1972 co-authored *Limits to Growth*—the book that sold millions around the world and launched a fierce debate on whether, and when, global consumption would overshoot our planet's resources. We've taught futures thinking together at the Norwegian Business School for more than a decade, and have also run the Center for Climate Strategies together. After decades of watching the world fail to take meaningful action on key issues, especially climate change, Randers recently wrote up his thinking about the most likely global future in the bestseller *2052: A Global Forecast for the Next Forty Years*. In it, he argues that rich countries will change their current course too slowly to avert severe climate disruptions. Innovation will lose to inertia. Humanity will hence fail the challenge to act on climate change before it reaches runaway proportions, in the main because people, capitalism, and democracy are too short-term to tackle the critical long-term climate issue.

Business as usual, he predicts, will run its course, overshooting ecological limits toward 2052 and beyond, but with better resource efficiency and less frantic economic growth than we've seen in the previous four decades. Randers forecasts that by 2100 we'll end up in a world three degrees Celsius (five degrees Fahrenheit) warmer. There'll be no end to setbacks and troubles, but we'll avoid the apocalypse of runaway climate tipping points at least in this century.[4] His story details a gray, muddling-through world, with slowing economic growth and mounting costs and consequences from climate impacts. Many, many natural species, habitats, and cultures, he worries, will gradually be lost, with the poorest people suffering the most.

The question that drives me is: Is humanity up to the task? Are we humans inescapably locked into short-termism? Both Randers and I agree that we already have the necessary technical solutions for a low-emission society. But he feels sure that our thinking is too short-term and our behavior too self-interested to turn around rapidly enough to avert runaway climate change in the coming century. Like Randers, many today argue that humans seem hardwired to self-destruct and take the planet's biological wealth down with us. Most do not even *want* to hear bad climate news.

I'd like to think otherwise, hence this book is a guide for how to break free from such a future forecast. It looks deep into the psychology of the human response to climate change and shows how to bypass the psychological barriers to action. Can those barriers really be bypassed, and soon enough to matter? Randers and I have been arguing about this for years, often in front of our classes. Where Randers, the elderly wise physicist, sees the socioeconomic juggernaut run its inevitable course of overshoot and decline, Stoknes, the younger optimistic psychologist, maintains that there is more to humans and cultures than short-termism. I argue—to his amusement—that the climate paradox is resolvable, and that solutions are within reach. He doesn't believe my scenario. I contest his forecast. So let's get going and prove my dear colleague wrong.

## Confronting the Climate Paradox

The climate paradox is easily evident. The scientific data and measurements about climate change and global warming are getting stronger and stronger. It's not that scientists are alarmists—it's that the science itself is alarming.[5] Still, people in many countries seem to care less and less—particularly in wealthy petroleum-based economies such as the United States, Canada, Australia, and Norway.[6] Heat waves are getting more frequent, stronger superstorms and typhoons are wreaking havoc in coastal settlements, sea levels are rising, the Arctic permafrost is melting faster than expected, corals and fish are dying, and there are more floods and droughts. I could continue, going through the whole usual list you will find in the fact-based scientific climate information, but I prefer to cut the litany short. Politicians have said for decades that they are concerned, and that "the time for action is now."[7] But talk is cheap, and there has been little decisive action and even fewer results.

In short, we know more than ever about this issue, and the situation looks graver than we thought. The technological part of the transition promises to be the easiest. We have the solutions we need to fix climate change: from radical energy efficiency to renewable energy, better education for women, reforestation, and carbon capture. Easy. But the public and political will is lacking. Reason has won the public argument about climate, but so far lost the case. Even if there is widespread concern, most Westerners still choose to look away—despite the dire facts, or perhaps exactly because of them.

Some of that apathy has its roots in deliberate denial and spin, but also to our susceptibility to it in the face of danger. So, part 1 of this book will take a detailed look into how the facts from the climate consensus are being shape-shifted into uncertainty, irrelevance, divisive fiction, hysteria, hoax, and conspiracy in the thinking of too many.

But even those who have tried to convey the alarming facts and moti-vate action have—often without realizing it—failed us. There is a golden rule in coaching and psychotherapeutic approaches to creating change: Our habitual solutions often become part of the problem. The standard response to difficult problems is to double our efforts. We try harder, pushing the old solution yet again. Being even cleverer at it, but getting more and more frustrated when the results don't turn out differently. Some deeply ingrained solutions are hard to unlearn. The solutions pushed by many environmental organizations have become part of the problem.

This is very evident in most attempts to communicate climate science to the public. When people aren't convinced by hearing the scientific facts of climate change, then the facts have been repeated and multiplied. Or shouted in a louder voice. Or with more pictures of drowning polar bears, still-bleaker facts, even more studies.[8] Still no response? Then the rule of thumb has been to try to shout louder yet. Make a hair-raising video with emotive music showing that we're heading for the cliff. Or write the umpteenth report for widespread distribution with the new facts, unequiv-ocal documentation, and lots of graphs, scientific references, and tables. Some are still surprised—or arrogantly annoyed—at all those people who just don't get it.

On the solutions front, carbon prices have been a favorite. We must raise $CO_2$ taxes, increase emission quota prices, and so on. In a perfect economic world with perfect markets, the Solution with capital $S$ is no doubt to set the right global cap and then the right cost for greenhouse gas emissions. When producing industrial goods in a global world, the associated "bads" of emissions should be taxed in an efficient manner. The polluter should pay. Both politicians and voters should support the carbon tax.

However, there is no global right price that all governments can agree on across cultures and local economies, despite the economic model saying this is the ideal. Neither is there one fair model for sharing emission rights among countries. Blaming politicians for these shortcomings doesn't bring much progress. Nor are there any institutions that can design and maintain the frameworks needed for this global top-down pricing and enforcement system to work.[9] Just assuming there ought to be and arguing from that assumption, as many economists have been doing, is acting like Peter Pan: believing that because we want it to happen, it should happen. Psychologists can recognize this as a form of wishful thinking, which is common not just among children, but among adults, too. But Peter Pan miracles only work in Neverland. Not here on earth.

For too long we've relied solely on this double push: More facts will finally convince the wayward about climate change. And there must be a global price on carbon emissions. Both are highly rational and uttered with the best of intentions, but neither is rooted in messy social reality or guided by how our brains actually think.

Don't get me wrong. I love rationality and clarity. I give offerings at the altar of Apollo, god of reason, logic, and academia. I'm not in any sense opposed to better facts, more rational communication, or higher prices on emissions. But rationality unfortunately has its limits. I wish as much as anyone concerned with the future of two-leggeds and more-than-human lives that this double push, gloriously rational as it is, would have been sufficient. Then it would already have solved our common problem. Yet it has repeatedly failed. Frustration, despair, and apathy have been the psychological outcomes.

The alternative to continuing pushing what doesn't work is: Try something else! And do more of what actually does work. There are hundreds of inspiring cases and examples out there that are already happening without waiting for a top-level climate treaty or high carbon price. Therefore, in part 2, I will review new examples and emerging strategies. These are things that are proven to work, based on human nature as described by psychologists and other social scientists. We'll explore how to use the power of social networks and norms, frame the climate issues in more supportive metaphors that avoid emotional backfiring, make taking action simpler and more convenient, and make better use of storytelling. Signals that give feedback on our progress are vital, too.

Luckily there is no need to shift everyone in modern democracies. Typically around 40 to 60 percent of people are already concerned, and politicians in principle only need a majority to press through stronger measures.[10] The challenge now is how to convert the felt concern into prioritizing the climate issue relative to other issues. Roughly one or two in ten need to shift into giving greater priority to ambitious climate policies. That would create a voter majority in favor of a great swerve.[11]

If part 1 explores how we *think* about climate and part 2 explores what new things we can *do*, then part 3 explores how we choose to *be* in the world. We (Western) humans have long understood ourselves and our economic systems as *separate* from the air, clouds, soils, rivers, and waters. The climate crisis seems to be forcing a slowly dawning recognition that we're intricately and intimately woven in with air, land, and sea. That lungs and leaves go together. And it's not just on the large, societal level. Where does your own self end and where does the air or water begin? Is the air inside my lungs *me*? The oxygen in my blood? Is the air I exhale *not me*? Is the water I drink, that comes into my cells, not me? The bacteria in my gut or on my skin?

And what about the village, town, or mega-city in which I live? We say, "I'm a New Yorker." My "self" is a web of relations. As humans, we are permeable to water, air, and food from the land. We're intimately connected to earthly flows, our technologies as well as more-than-human beings in myriad ways. So part 3 questions whether we can continue to regard the world as out-there and separate from us, and shows how to make peace with the restless, living air.

—PART I—

# Thinking

## UNDERSTANDING THE CLIMATE PARADOX

*We see and hear what we are open to noticing.*
*—JEROME BERNSTEIN*

# ONE

# The Psychological Climate Paradox

We know that climate science facts are getting more solidly documented and disturbing year by year.[1] We also know that most people either don't believe in or do not act upon those facts.

It forces the simple question: *Why?*

All science academies in the world have double-checked the climate facts and the models.[2] Skeptics have triple-checked them. Even skeptics previously funded by the petroleum industry, such as Berkeley Earth, are now convinced.[3] The science is settled on the question of *whether* human-made global warming is happening. This book will not review that science, since that has been covered repeatedly elsewhere.[4] The remaining scientific questions are those of how rapidly? How much? What, where, and how strong will the local impacts be as we crawl through the century? These questions, too, are outside my scope. This book's core question is rather *how people respond* to such messages, not the facts themselves. Facts and beliefs are very different things. If new facts don't support a person's beliefs, often "the facts bounce off."[5]

To communicate the facts to the public, the same social experiment has been repeated over and over: Simply give people the information, and then wait and see if the facts trickling into their mind will convince them to change their behavior. The outcome has been consistently underwhelming.[6] But that hasn't held rational people like climate scientists, public servants, and environmentalists back from trying the same experiment on the public again and again—each time with yet more facts and, each time, for some weird reason, expecting a different outcome. But inadvertently, conventional climate communications have triggered more distancing, not increased concern and priority.

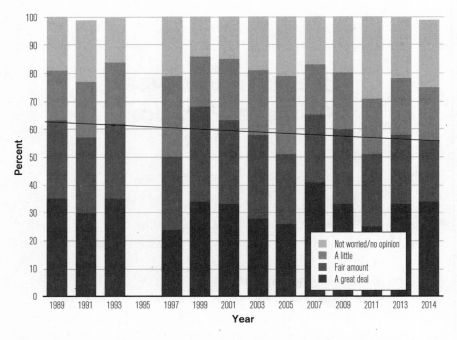

**Figure 1.1.** US responses to the question, "How much do you personally worry about the greenhouse effect or global warming?" Source: Gallup, 2014.

So climate change enlightenment was fun while it lasted, as George Monbiot noted.[7] But it is now limp and dead.

So dead, in fact, that it is moving backward. Surprisingly, the level of concern among both laypeople and politicians has actually been *decreasing*—especially in many wealthy countries—over the last two decades.[8] This is the opposite of what you'd expect from scientific reasoning. As figure 1.1 shows, in the United States concern about climate change has weakened overall since 1989, despite that the objective data have been strengthened with thousands of studies and reports. The same holds true for other rich nations. Norway, a rich oil country, shows an even stronger decline in concern (see figure 1.2). Some psychological studies even point to a strange relationship between global warming denial and speaking English in particular, since the United States, UK, and Australia are countries with waning levels of average public concern.[9]

In a 2013 study, the Pew Research Center asked people around the world to rate their concern over climate change, financial instability, and

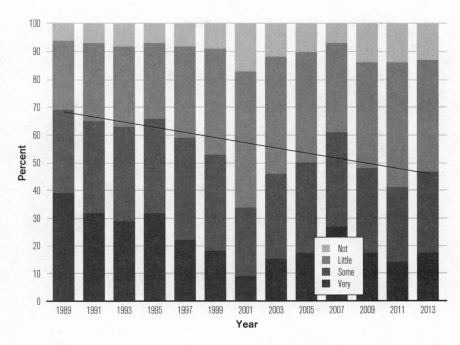

**Figure 1.2.** Norwegian responses to the question, "How concerned are you about greenhouse effects and climate change?" Source: Ipsos MMI, 2014.

Islamic extremism, among other topics. For US residents, concern over climate ranked lowest (see figure 1.3), while concern about climate ranked highest for the public in developing countries.

This weird state of affairs calls for a radical rethinking in how to communicate climate change. It seems that for most people in rich countries, the climate message of increasing disaster, damage, and doom is uncomfortable to live with. It tells us that we are partly at fault for the destruction of the planet. Further, the scientific message and the recommended policies (carbon taxes, regulations, less travel) are not just inconvenient, but highly unpalatable, almost unspeakable in some circles. These awkward feelings must then be done away with.

One psychological reaction might be to deprioritize the issue. And that is what many do. After all, it's not that there is no concern about climate change. A clear majority (54 percent) is concerned across all countries.[10] But more tangible worries such as job security, financial turmoil, school quality, and health have all reduced the priority for climate in Western

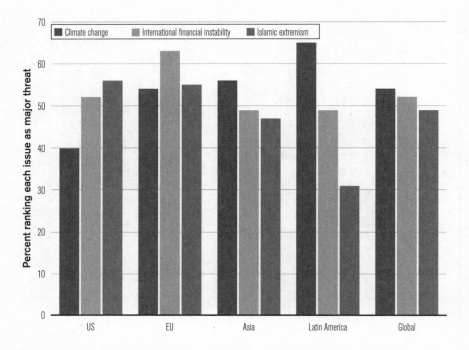

**Figure 1.3.** Global and regional views on major threats to their country. The graph shows the responses to the question, "I'd like your opinion about some possible international concerns for [survey country]. Do you think that [global climate change] is a major threat, a minor threat, or not a threat to [survey country]?" Source: Pew Research Center, 2013.

societies over the past years. Faced with nearer concerns, the worries for climate are displaced. As their top environmental concerns, Asian Pacific consumers cite water shortages and local air pollution. North Americans and Europeans cite water pollution and shortages.[11] When ranking the top public overall policy concerns, the climate typically gets even lower ratings, like nineteenth out of twenty on the list, way below strengthening the economy, jobs, terrorism, and education. So despite the general climate worry, when prioritizing the concerns, climate loses out both as a top policy concern and among other environmental concerns.[12]

Others take it a step—or many steps—farther than just low priority. Faced with the psychological burden of climate information, they decide the science itself must be false. Very creative psychological solutions are quite easy to come up with in order to dispel the climate science. Many therefore prefer to label climate change as natural: "It's always changing."

And they decide to view global warming as an exaggerated media story. If you can convince yourself it is not important, it can then be dismissed. Others ridicule it as a hoax, or a conspiracy against free thought, free choice, and free enterprise. They might say that the scientific consensus just displays groupthink among a greedy, closely knit power-hungry sect calling themselves scientists. Some call it a new irrational religion. Neo-Marxist propaganda. A trick to get more research money. And a threat to our Western way of life, which, as George H. W. Bush famously declared, "is not up for negotiation." All of this, and more, has been floated in the wide world of denial.

How and why did climate science turn into a politically alienating issue? Clearly, if there ever was a common cause for humanity, a shared vision of the put-a-man-on-the-moon type, then working together to maintain a stable and friendly climate for our societies now and into the future could or should be it. How and why has such a potentially unifying vision turned out increasingly divisive?

Now, paradoxes are neither new nor rare. This is not the first time humans have acted paradoxically. We talk about peace and go to war. We love our children, yet hurt them. We want to be slim, yet overeat. We work out to be healthy, yet smoke and drink too much later the same day. We praise the wilderness, and cut it down. We talk about the long term, and act as if only the short term matters. In other words, we two-leggeds are predictably irrational.

So the point here is not to flagellate and criticize our all-too-human selves for not being consistent, for not infallibly speaking with one voice, for not meekly obeying the rational facts. Rather, double standards are the rule. We pretend to be rational while behaving irrationally. No lack of paradox in the human psyche. And we shouldn't expect there to be. Therefore, it is now time to have a more generous, empathic look at our climate irrationality—as well as at how this has been exploited by certain interests.

We need a more compassionate understanding of how our paradoxical, many-voiced psyche is responding to news of the ongoing shifts in the climate. We're suddenly face-to-face with the responses of the air itself to human emissions. The air is shifting its ways since we've changed its makeup, shifting the way it flows, murmurs, and roars, its own Logos. At some level we are aware we're causing this shift ourselves. We know we've done something irreversible. Just like Adam and Eve, we've eaten the prohibited apple. Only this one is a carbon apple, the black fruit from ancient trees in the prehistoric gardens. Yet we pretend we're innocent. Or that we didn't know what we did. As if we weren't black all around our insatiable mouths. We naïvely expect we can dump billions and billions of tons of gaseous pollution into this Eden, this thin air skin of the earth, this highly complex and delicate weather dance, and there will be, should be, no consequences at all. It truly is a weird, wide world of denial out there.

# TWO

# "Climate Is the New Marx": The Many Faces of Skepticism and Denial

On a personal note: I am a scientific skeptic, and thus a climate skeptic. I've spent most of my professional life consulting with and teaching about how to deal with uncertainties in our future, specializing in scenario methods.[1] Of course, the numbers used in future climate change scenarios are uncertain. Science always deals with estimates and probabilities, not with absolutes. The climate models are fed with approximate data and then calculated with error bars. The temperature projections for the coming century and centuries contain substantial uncertainty. The skeptic attitude is essential to the scientific process itself. Good climate scientists and peer reviewers must be critical and skeptical about both the input and output of experiments, measurements and models. Skepticism as philosophy says there is no such thing as absolute scientific proof of any statement about the world.

But skepticism has its limits in practice. Uncertainties pertain to some areas more than others. Some knowledge is certain *enough*, even if it might *in principle* be proved false in the future. For instance, while you are sleeping tonight, somebody *may* remove the floor beside your bed. So when you step out of your bed, you may fall many meters. A radical skeptic must remember to check that the floor is still there each time he gets out of bed. Or that the tires on his car haven't been removed while he was shopping. And check the brakes and . . . tiresome, isn't it?[2]

The limits to legitimate skepticism may best be described by the phrase *beyond reasonable doubt*. It has to do with a clear balance of evidence. Among scientific certainties that are beyond reasonable doubt are: that the earth revolves around the sun, that natural selection has formed species, that DNA is a form of code for cellular proteins, that excessive smoking increases lung cancer, that DDT kills off insects until they develop resistance, that $E = mc^2$, that increasing $CO_2$ levels in the atmosphere trap heat, that trapping solar heat leads to higher average temperatures in air, ocean, and soils over time.

It is a good thing for all of us that some thinkers and scientists take the effort to seriously question and audit every one of such commonly accepted certainties from time to time. Maybe we got something wrong some time back. If someone is convinced there are fundamental errors, and puts in serious efforts to research and document findings that support that contrarian claim, those efforts are invaluable to science. Consensus around the dominant paradigm needs deep questioning from time to time to avoid becoming too stale and dogmatic. This is part and parcel of the scientific method and process, indeed key to its developement.[3] Still, if the supporting data is massive at a point in time, then this evidence is beyond reasonable doubt. Such facts—or even "truths"—are solid enough to be acted upon, while a few contrarians keep up their valuable, critical work.

We should also remember that there is lots of skepticism voiced among the mainstream climate scientists. They question and quarrel over one another's results and models at great lengths. They are skeptics who think critically about results, and question assumptions for projections. Science itself is really systematized skepticism. Thus, being a skeptic is, in science, a generally positive label. What often passes as a climate skeptic in media parlance, however, is an altogether different type of cake.

The media has transferred the label to people who consistently *choose to believe* in the *opposite* of carefully critiqued climate science evidence. That choice may result from what psychologists call *denial*, a term that unfortunately is too often used as a pejorative or insult. The point is that some people are unflinchingly convinced that climate science is wrong or unsettled. This is not a skeptic position, but one of faith. Often these people even use the rhetoric of skepticism to cleverly cover up an underlying fundamentalist stance.

A genuine scientific skeptic would say something like this: "Based on the balance of evidence we have up till today, human-made global warming seems grave and very likely. But you can't rule out that *maybe*, next year or next decade, we might have even better knowledge, either way." Few things would make me happier than if it turns out, ten or twenty years from now, that $CO_2$ emissions aren't as destructive as the evidence shows today. That would buy us a little more time to respond. But I remain highly skeptical about that possibility, since the heating and acidification have been demonstrated beyond reasonable doubt. Still, it cannot be *completely* ruled out. We cannot become *believers* in a certain climate scenario, even if the balance of evidence is very clear today.

The climate deniers, however, are those who say they are convinced the whole thing is overblown, wrong, a hoax. Or a conspiracy. And they use ridicule and scorn to argue their viewpoint. So while most media mixes skeptics and deniers, in this book I'll reserve the term *climate skeptics* for those with genuine skepticism. I'll use *contrarians* for the few who hold skeptical, rational minority positions,[4] and *climate deniers* for those with a belief-based and anti-skeptic attitude. A detailed exposition of the psychological dynamics underlying this type of denial is—sadly—very much needed at this time.

I have been collecting 'species' of denial for years in my home office, like others collect stamps or coins for curiosity and learning. I think it is important to see and study the actual wording and reasoning in some detail, so that we can get a feel for how the mind moves. If you feel that the quotes below become tedious, jump to the next section. I think the following selections are intriguing specimens:[5]

- One Norwegian populist politician, Per-Willy Amundsen, explained his position to a major newspaper: "Some climate fundamentalists see nothing but climate change in every hurricane or extreme weather event. People now believe they have to move up in the mountains if the Greenland ice melts . . . [but] Karl Marx is dead. Socialism is defeated. To replace Marxist theories, the left are now picking up climate theories. For them, this is all about finding new excuses for taxing and controlling the people. Previously it was capitalism that was destroying the world. Now it is the climate. For socialists adrift with no proper anchor, climate theory is now the best to believe in."[6]
- The US public policy pundit Pat Michaels used his *Forbes* column in 2010 and 2011 to claim that there hasn't been "much warming since the late 1990s," a claim he makes repeatedly.[7] He is joined at *Forbes* by Larry Bell, a professor of architecture, who claims there is a "feverish hunt for evidence of a man-made global warming crisis" because the planet

is actually cooling, and that "continued cooling may jeopardize climate science and green energy funding."[8] The same Larry Bell also published a book in 2011 on the "global warming hoax," intended to bring "welcome relief to all those who are fed up with climate crisis insanity."[9]

- Kentucky state representative Kevin Sinnette, a Democrat, holds the opinion in 2014 that climate change didn't kill the dinosaurs, so human beings should be just fine now: "The dinosaurs died, and we don't know why, but the world adjusted. And to say that this is what's going to cause detriment to people, I just don't think it's out there."[10]

The US House of Representatives has a pretty outstanding record of climate denials, from which I'll quote quite a number (though there are many, many more to pick from):[11]

- Chairman Emeritus Joe Barton (R-TX) stated that "the science is not settled and the science is actually going the other way."
- Ed Whitfield (R-KY), the chairman of the Subcommittee on Energy and Power, called on Al Gore to "come clean about the real science surrounding climate change and let the American people come to their own conclusions on global warming."
- John Shimkus (R-IL), the chairman of the Subcommittee on Environment and the Economy, rejected the dire warnings of climate scientists and said the earth "will end only when God declares it is time to be over. Man will not destroy this earth. This earth will not be destroyed by a flood."
- John Sullivan (R-OK), vice chair of the Subcommittee on Energy and Power, stated, "I don't think anyone could come to any conclusion whether it is real or not. Until we can see sound science that's truthful, I don't think anyone can make a decision based on that."
- Michael Burgess (R-TX) stated that "no one knows" whether humankind is responsible for climate change. He said it is "just the height of chutzpah for us to be claiming that man made effects can change something as profound as the climate on this planet. The climate has changed over eons. Man has had nothing to do with it."
- Morgan Griffith (R-VA) called it "reckless" to cut greenhouse gas emissions "in order to address a scientific theory—man-made global warming—that many scientists do not even believe is happening."

- At a press conference in connection with the Doha December 2012 climate negotiations, US senator James Inhofe (R-OK) said, "It's time to put an end to these lavish, absurd global warming conferences and focus on the real problems that we face as a country. The focus of this year's global warming conference, like all the conferences before, is not about the environment. It's about one thing: spreading the wealth around."[12] Inhofe also said, in 2003, "With all of the hysteria, all of the fear, all of the phony science, could it be that man-made global warming is the greatest hoax ever perpetrated on the American people? It sure sounds like it."[13]

- A US think tank called Committee for a Constructive Tomorrow released a report, *Extreme Weather 2012*, that "debunks claims that severe weather is becoming more frequent, a common assertion of global warming activists in the wake of Hurricane Sandy." The author, Marc Morano, said his report "reveals that the latest peer-reviewed studies, data and analysis undermine the case that the weather is more 'extreme' or 'unprecedented.' On every key measure, claims of extreme weather in our current climate fail to hold up to scrutiny."[14]

- Lord Christopher Monckton, another prominent self-proclaimed climate skeptic, writes: "For years the true-believers had gotten away with pretending that 'climate deniers'—their hate-speech term for anyone who applies the scientific method to the climate question—do not accept the basic science behind the greenhouse theory . . . The IPCC never had a useful or legitimate scientific purpose. It was founded for purely political and not scientific reasons. It was flawed. It has failed. Time to sweep it away. It does not even deserve a place in the history books, except as a warning against the globalization of groupthink, and of government."[15]

- A *Washington Times* editorial states: "Even though the official charts show no significant warming trend in the past 15 years, the planet may be even cooler than the IPCC figures suggest. [Meteorologist] Mr. Watts, who runs the Watts Up With That website, points out that IPCC is using adjusted data. In a forthcoming scientific paper, he demonstrates that improper placement of weather stations has resulted in the temperature increase being overstated by 92 percent. The last thing government officials want to hear is that the planet isn't actually warming."[16]

- In Australia, Alan Moran, an economist and director of the Institute of Public Affairs Deregulation Unit, writes: "Alarmist stories about greenhouse gases causing catastrophic warming continue to be aired in the media. Notwithstanding the lack of evidence, global warming is even being blamed for hurricanes and an apocryphal disappearance of polar bears. Yet, the only solid measure of the warming, the NASA satellite data, shows that over the 27 years that data has been available, warming has been at a negligible rate of 0.13 degrees Celsius per decade."[17]
- Alex Jones's website infowar.com wrote in September 2013, after the first release of IPCC's *5th Assessment Report*: "Even the latest UN Intergovernmental Panel on Climate Change (IPCC) report seems to indicate that an era of global cooling is now underway, according to many scientists. It turns out that global warming predictions were little more than doom-and-gloom fear mongering based on failed computer models. For example, in 2007, the BBC reported that the Arctic would be 'ice-free' by the summer of 2013."

The BBC article did not say "would" but "could."[18] Sometimes shifting just one letter makes a lot of difference. The BBC article also contained the statement, "It might not be as early as 2013 but it will be soon, much earlier than 2040."

There is a true abundance of such quotes, but I won't indulge in more. They crop up particularly in unedited blogs, think-tank-funded publications, and commentaries on the Internet, where many repeat and refer to one another. And in the mass media, these active denials are often quoted by journalists who want to give "both sides" a voice, which in practice means that a few prominent trigger-happy deniers get proportionally much more media coverage than the 98 percent majority of skeptical yet concerned climate scientists.[19] Anyone can look up a climate science news article in online news media. Then scroll down and have a look at the comments section, where readers can voice their opinions. You will always find someone eager to tear apart the article, the messenger, and the science. I picked these two, more or less at random:

The problem with AGW [anthropogenic global warming], which is essentially a religious type cult, is to weave a narrative of the world for the believers. You may be losing control of publications—old timers are still sure that nothing anti AGW will ever make it through peer review . . . People who believe that less Arctic ice in September causes an Oklahoma storm next May will believe in EVERYTHING. In everything, yes, as long as there is SOMETHING, no matter how far fetched.[20]

In this one, from WattsUpWithThat, the anonymous commenter Cheshirered calls for responsibility:

It is surely time for a collective call to account to be issued—by a single body representing all sceptic organisations, scientists, journalists and blogs, to those responsible for this nonsense: the IPCC, leading alarmists & NGO's, the national governments of the US, UK, and the EU and to the alarmist media. It's time they were held to account for wrongly promoting what are now transparently failed projections, improbable scenarios and impossible outcomes. AGW theory is dying on the vine right before our eyes. There is not one key indicator that is falling in favor of AGW theory. It's way past time those responsible for pushing this junk are held responsible.[21]

In these last two quotes, gleams of paranoid reasoning, threats, and ardent anger spill through the language. There are many, many, many more such comments out there.[22] It is clearly the case that climate science results generate a lot of heated resistance and opposition.[23] The climate issue seems to attract a lot of discontent, and works as a projection screen onto which it can be vented.

## States of Denial

What makes the above examples cases of denial? Are they not just free speech, ordinary and acceptable differences in opinion? One says this, another says that. Everyone is—in democracies and on the Internet—entitled to their own opinion. But everyone is not entitled to decide upon and make up their own *scientific* facts. You may be free to state that you don't like some scientific results. But not to twist or label "junk" or "pseudo-science" those you don't like. Nor to communicate cherry-picked subsets of facts in any way that pleases you.

On the other hand, razor-sharp critique and contra arguments are not just welcome, but necessary. It is good that people want to independently audit scientific results and demand access to the raw data. We must welcome engagement for better transparency and repeatability of scientific methods. There is always much to correct and improve in science. Scientists make mistakes all the time. That's why peer-review and skeptic arguments are vital.

But there is a vast difference between putting forth a point of critical view, honestly held and well argued, and intentionally sowing the seeds of confusion in an angry rant because one feels affronted by scientific results. Sometimes the line is subtle and hard to spot. But usually, if you look at the manner and style of speaking, then the red temper, capitalization of letters, sarcasm, and ridicule employed unwittingly expose the underlying motivation. It is not to promote a free, skeptical, tough-minded view, but to mock and discredit the "other" side. Free speech does not grant a license to deceive, shame, or threaten.

Denial was originally a psychological concept, influenced in particular by the founder of psychoanalysis, Sigmund Freud. He described it as an automatic inner infantile defense mechanism that protects the ego, or our "I," against perceived threats. This mechanism kicks in automatically to avoid the feelings of fear, anxiety, or hostility, in particular those arising from inside our own unconscious.[24] Since Freud's time, our understanding of denial has undergone a long series of updates and revisions.

My working definition of denial follows sociology professor Stanley Cohen's thorough review in the classic book *States of Denial*. Cohen studies how people and governments choose to ignore and deny facts about atrocities and suffering, from apartheid to Palestine. He asserts that denial stems from *the need to be innocent about a troubling recognition*. It is both to know and simultaneously not-to-know. It is to be aware of something, yet at the same time argue and present its opposite in a convincing way.

Some say that denial is lying so hard you really come to believe in your own lie. This is usually done to justify your own position in the face of the troubling fact or recognition.

*Active* denial is somewhat distinct from *passive* denial.[25] In active denial, there is knowledge of the facts to a certain degree, but they are energetically refuted and refused and rewritten since they don't fit with your values—personal, political, or otherwise. An American officer said after the My Lai massacre in the Vietnam War: "There was no massacre and the bastards got what they deserved."[26] In this example what is refuted is also indirectly accepted in the very same sentence. This happens when the knowledge you hold would, if acknowledged and acted upon, threaten your income, profession, or status. Or it may disturb your thinking and self-image, and thus be psychologically painful to acknowledge. In the same way the series of quotes above are examples of active denials, not genuine skepticism.

*Passive* denial is more on the indifferent, unresponsive side. You may know climate change exists, but prefer not to care much about it. Or you might employ some irony and sarcasm to fend off the subject when it is raised. But sometimes the silence speaks volumes. The issue becomes unspeakable. You may even care, be worried or in despair over the droughts, heat waves, and loss of habitat, but fear or learned helplessness keeps you from acting and reacting: "There's nothing I can do about it!" From there, sliding into silent disbelief offers itself as a convenient way out of the discomfort: "I can't do anything about it, therefore it is not real." Best not to mention it at all.

In order to understand how people are living in this passive denial, sociologist Kari Marie Norgaard, from Oregon, interviewed inhabitants of a Norwegian village that was experiencing a period of warm, weird winters. One person held his hands in front of his eyes and said, "People want to protect themselves a bit." A student observed, "Despite my knowledge of the wider climate issues, I am still living the same life." Another man expressed a sense of concern about future impacts: "I see that we do lots of things that most certainly cannot continue. It will work for a while, but sooner or later it isn't going to work. So I am worried in any case for that which will happen."[27]

Yet another said, "I don't completely know what I shall think of it. But regardless, I believe that many believe that it's wrong that we are changing nature so much. I have the sense that most people believe that nature knows best. I believe that. So, basically, I think that people are worried about climate change. They don't know rightly what to say."[28]

There is always a component to denial that goes deeper than likes or dislikes. You're not in denial about roses if you don't like roses, but prefer tulips. Or about politics if you can't stand a certain politician. Or if you want to switch off the radio each time pop artist Justin Bieber sings "Baby baby." The concept of denial is reserved for those issues that are emotionally and morally disturbing and therefore—if not dealt with—generate an uncomfortable inner splitting. The term *denial* is appropriate when the full acknowledgment of what is denied would imply *having to* act upon it. The lifting of denial would result in an emotional shift, and would require both speaking and acting differently. And sometimes it would result in a substantial change of lifestyle, ethics, and identity.

To what extent this individual shift out of denial is needed for cultural shifts to happen, and what effects such individual shifts would have, is shrouded in uncertainty. For instance, many whites in South Africa in the 1980s were in denial about apartheid. If asked, most would defend the notion that it was right and necessary to keep blacks segregated. Leaving that position and speaking out against it could come at a great cost to their standing at work and with the authorities, family, and friends.[29] Still, the cultural shift happened in the 1990s. The same type of shift out of cultural denial preceded the abolition of slavery during the nineteenth century, changed the UK attitude toward Hitler in 1939 and the US attitude toward the Vietnam War in 1970, spurred the fall of the Berlin Wall in 1989, and got people to acknowledge the effects of tobacco smoking in the 1990s.

In addition to active versus passive, and individual versus cultural denial, there is also *denialism*. This refers to a certain style of rhetoric. It gives an appearance of legitimate discussion, when in actuality there is hardly any data or proper argument. Its goal is simply to dismiss the scientific consensus. False arguments are built on cherry-picked facts and dressed up in forceful expressions to bolster someone's personal viewpoint against overwhelming evidence: "It has actually been cooling since 1998!!!" This may be entertaining on TV shows and successful in distracting from useful discussions—but only by using appealing yet ultimately empty and illogical assertions.

"Don't mistake denialism for debate," says Mark Hoofnagle.[30] He points out that denialism uses five typical tactics that gives it a recognizable profile:

1. The identification of conspiracies, which is an attempt to take the man, not the ball.
2. The use of fake experts. This can be using or paying a prominent expert in another field to lend credibility.

3. Selectivity—drawing on isolated papers that challenge the dominant consensus or highlighting the flaws in one or two of the weakest papers as a means of discrediting the entire field.
4. The creation of impossible expectations of what evidence research can deliver—a perverted form of skepticism.
5. The use of false analogies and logical fallacies. For instance, when the EPA said that tobacco smoke was carcinogenic, this was described by two commentators as an "attempt to institutionalize a particular irrational view of the world as the only legitimate perspective, and to replace rationality with dogma as the legitimate basis of public policy." They claimed it was a "threat to the very core of democratic values and democratic public policy."[31]

"Denialism is pretty predictable and consistent in form no matter what the topic," concludes Hoofnagle.[32]

Can reversing cultural and psychological climate denial be thought of as a cultural transition on par with the transformations that led to the abolishment of slavery or apartheid, the end of the Vietnam War, the widespread rejection of smoking, or the dissolving of the communist system in Eastern Europe? I claim it is similar, and therefore it is worth studying. The big question, of course, is: How long until we'll see a similar seismic shift of cultural climate denial in Western democracies?

Psychiatrist Robert Jay Lifton calls this a swerve in cultural awareness. In the 1980s people came to feel that it was deeply wrong, perhaps evil, to engage in nuclear war. Now, he says, they are slowly "coming to an awareness that it is deeply wrong, perhaps evil, to destroy our habitat and create a legacy of suffering for our children."[33]

Psychology claims that denial *can* be dissolved through a process of insight and acknowledgment. For some, such a change in worldview will never happen. Such persons bring their convictions stubbornly to the grave. For others, there may be a gradual—or sudden—realization that one's self-understanding and behavior have to change. Which comes with a certain level of pain and possibly a huge social cost. You might even have to change your friends and social network.[34] Or some later reframe it so that they can change position, yet claim that "this is what I meant all the time." But if we persist in acknowledging denial, then both personal and cultural shifts can reinforce each other, making way for societal transformation. It has happened before, on all kinds of topics that our cultures have been in denial about.

## At What Levels Is Denial Really Holding Us Back?

Is denial as a psychological concept applicable beyond the individual? Does it influence what is going on even at the national level? Or international? Perhaps so, but many social scientists prefer to avoid the denial issue and skip the individual level of analysis. To understand why so little progress is being made, they say, answers should rather be sought on the international, national, organizational, and cultural levels.

Internationally little happens, political scientists argue, because countries choose to compete and disagree rather than cooperate. If the world's countries were acting for the common good, they would quickly agree on a global price of carbon—and stark regulations and enforcement procedures to go with it. However, most countries seem more eager to grind their own mills. The poorer nations want to grow faster, and the old industrialized countries want energy security and continued growth. The oil-rich countries want to continue to sell their black gold.

Each nation is acting rationally if seen from a narrow self-interested point of view. But take a look at international climate summits. Even if there's no end to all the good words and phrases and intentions, by the end of each event the only thing everyone can agree on is to have another meeting next year. Participants' self-interest works destructively on the common level. What's rational for each country by itself creates collective madness.

According to this view, countries march collectively toward the "tragedy of the commons."[35] That means moving inevitably toward overall climate disruption of unknown extent. Even if no single country wants that to happen. The mad logic is: *Better that my nation-state doesn't lose in the short-term race than for all of us to win in the long term*. No delegation representative wants to come home and explain to his or her citizens that they have committed to more seemingly expensive obligations than other countries have.

But if international negotiations don't work well, then surely at the *level of countries*, each nation-state can do something. The national governments are powerful, right?

Well, most governments see their primary task as maximizing the wealth and welfare of their own citizens, not other countries' citizens.[36] And with no global price or binding treaties, there is little to be gained by setting higher carbon taxes or more ambitious regulations than other countries do. Any governments that do so harvest mostly criticism from interest groups and voters. Also, public economists have been clever at arguing that it is not cost-efficient for any single country to do more than others. If one country were to set carbon taxes higher than others, its economists would fear a loss of competitiveness and industry. There are cheaper emissions cuts to be done abroad first, they say, before cutting at home.

Public servants have no incentive in advocating for costly measures or risky climate-related investments. They don't get rewarded for recommending risky new green technologies even if they turn out successful, but their personal reputation will certainly suffer if it goes badly. So for public servants, there is no upside to being proactive, only downside. Finally, most national politicians are by now aware that the climate issue is critical in the long term and know—at some level—what they ought to do immediately to mitigate it. There is no lack of well-documented solutions, such as taxes, cap and trade, regulations, and subsidies for better technologies. But climate policies that really cut emissions drastically, such as doubling national taxes on energy use or emissions unilaterally, have neither popular support nor industry support, nor are they deemed cost-efficient. Politicians, once in power, may know what they ought to do, but not how to get reelected by the citizens after they've done it.

Then there is the *organizational or business level*. Industry and corporations have the capacity to cut emissions. So why don't they? Well, most executives believe their primary job is maximizing profits by focusing on their core competency and core business. But reducing emissions often means increasing upfront investments and operational costs in order to cut costs that will accrue years into the future. The climate issue is considered outside the core business. Since a high discount rate is used in calculations of the return on investment, management more often favors short-term marketing, upgraded products, or capacity utilization rather than future energy savings.

All the big oil companies claim, now, to be green or at least getting greener. ExxonMobil, Statoil, Shell, and China National Offshore Oil Company all have glossy sustainability annual reports. They say they use high implicit carbon prices to guide investment decisions. Nearly all claim to support international climate policies. They rarely state openly that they expect future climate regulations to fail.[37] And then they conveniently end up waiting for national and international action on higher carbon prices. Meanwhile, business as usual—exploration and drilling—is the most rational option for each of them. Even if they individually are well aware that added together, all their current reserves of fossil fuel are sufficient—if all extracted and burned—to utterly disrupt the earth's climate.[38]

At the *cultural level*, strong social norms support the status quo. It is not easy to be a climate alarmist alone at work, at school, or at home. When the rest of your friends and family are focused on their daily worries, who wants to speak about climate doom? People spend their daily lives thinking about more local, manageable topics, which are easier to talk about. A great silence surrounds climate change in everyday life. "What to pay attention to and what to ignore is socially constructed. We learn what to see and think about from the people around us," writes Norgaard.[39] If your sibling or in-laws work for the petroleum industry, then arguing over lunch or at Christmas parties that drilling must stop is not a winning topic. Most of us end up subtly adjusting our attitudes to those that our significant others express.[40]

So there are strong barriers at international, national, corporate, and cultural levels. Change on these levels is agonizingly slow, in light of what climate science tells us about the urgency to reduce emissions. More than twenty years of climate negotiations have resulted in almost no implemented reductions. Most of the larger reductions that have happened since 1990 have come about by other means, such as efficiency gains in energy or the collapse of the Soviet Union with its gigantic, inefficient industries. Global and national leadership is missing. Political scientists, economists, strategists, sociologists, and anthropologists each have their favorite levels to explain why we disagree, and why so little is happening.[41] None of them really address the role of denial.

That's why I'll look primarily into the more basic individual level, including the *psychology of denial*. But I'll also look at how ideological denialism plays itself out in the social and cultural networks. Since international, national, corporate, and cultural levels all seem pretty stuck, no major decisions will be made—at least in democratic countries—until vocal and numerous citizens demand it. Rather than just waiting for the top-down approaches, maybe we could start by understanding apathy and denial at the individual and small-group level. Then, through lifting denial and shifting social networks, we could see a growing bottom-up support for stronger measures and actions at cultural and national levels. Maybe millions of nongovernmental networks and organizations as well as thousands of green businesses and cities will drive the great swerve. This is my hope and purpose: that by understanding the barriers in our thinking (part 1), we may find new ways beyond them (part 2) and learn how to support the more collective levels from the bottom up toward a new way of being in the world (part 3).

## Denialists: Villains or Victims?

Is it even appropriate to put psychological labels such as *denial* on some global warming contrarians? In the Soviet Union under Stalin, people with deviating opinions—those who opposed Stalin's totalitarian and paranoid system—were often given psychiatric diagnoses.[42] Many were exiled to Siberia for a tortuous "cure." Some climate contrarians and deniers have been turning the Stalin argument against the scientific climate consensus, saying that people voicing disagreement against the consensus are censored and labeled as deviants. They claim there are active attempts to silence them from voicing their honest opinion against the mainstream. They see a dogmatic climate fundamentalism that strangles free thought. And that peer review of scientific journals works just like Stalinist censorship in stifling legitimate dissent.

In their own eyes, they are victims persecuted by the dogmatic majority. But they have the courage to stand up for what is now politically incorrect. Even if these self-proclaimed skeptics are many and everywhere (typically 30 to 60 percent of wealthy Western countries[43]), they still fancy a story about themselves as victims of suppression. The most active of them like to view themselves as the lone dissident voices, articulate clearheaded heroes, mavericks, and fearless guardians of the obscured and abused Truth. They liken themselves to Galileo fighting the stifling church. They have to struggle through a vast flock of sheepish yea-sayers being led astray by the wicked, self-interest-driven, and powerful climate lobby. But they will never, ever succumb to the AGW cult of the climate alarmists!

Is there something wacky about using psychoanalysis on the types of climate naysayers described above? Yes, to psychoanalyze people who disagree with your political opinions signals intellectual arrogance, or at best closed-mindedness. The use of psychiatric diagnoses as a power play has a long and very dark history, not just from Stalin's times.[44] However, when using psychological analysis on climate denial, my purpose is not to blame, pathologize, or suppress. Rather the aim is to facilitate a change process through more empathic understanding—as I attempted with the machine-industry executives in that hotel conference room. I am genuinely curious, sometimes both amused and saddened.

What bewilders me is why the intelligent minds of so many turn against climate—and not just among the most outspoken and die-hard deniers. How is that so many citizens have heard the facts, yet still resist or ignore them by choosing to lend an ear to outspoken and angry denialists? If it works at all, psychology works by approaching those who seem (at least at first) weird and self-destructive empathetically, to find that their experience is human, too. We stop blaming and shaming and listen intently to both conscious and unconscious motives. I agree that this approach can go too far: Taken to the extreme, the empathic psychologist would understand everything, embrace all, and forgive all. The purpose of psychoanalysis, though, is neither of these extremes—not condemnation and not pampering.

There is rather a deeply *ethical* issue at stake here.

Some issues don't lend themselves to indifference and free choice. It is not ethical to say: "Some like slavery, some don't. Everyone can have their own opinion." There are no circumstances that make brutal fourteen-hours-a-day labor for ten-year-old boys acceptable, for instance; no ethical ways to avert your eyes if you see someone beating and kicking the life out of another lying on the street. There is no tolerable tax haven, sex trafficking, rape, corruption, or torture for caged animals. The ethics philosopher Emmanuel Levinas says the ethics of such issues stem from the face-to-face immediate encounter with another. The suffering in the face of the Other limits your freedom to just pick any opinion at your whim and then use all your intelligence to back up and rationalize that position, polishing it until it seems unassailable and impenetrable.

These same ethical parameters arise on the climate frontier. There are serious risks that climate disruption poses to *us all* and particularly to the poor and to the more-than-human beings *that do not even participate in the climate debate.* They have no voice in opposing rich countries' climate emissions that slowly destroy their home, their ecology. Many others have no voice because they haven't been born yet. It can therefore no longer be a case of "anything goes."

I don't care much if you're pro tulips, hate pop star Justin Bieber, or are anti-gluten. Nor if you hate taxes, feel like outlawing financial speculation, or never want to see another fat-cat investment banker. But when it comes to the facts of increasing climate disruption due to our human impact on the earth, there is an ethical obligation to respond. Those who actively give voice to or just passively live out their denial actually support the ongoing human violence to the land and clean air. This includes the humans who breathe and live in these landscapes. If we support denialism, then we lend our cognitive capacity, creativity even, to support a destructive cause.

The stunning human potential for creativity and cooperation is matched only by our vast potential for self-destruction.[45] This potential was amply demonstrated in the previous century with its two massive hot wars and then a cold one. Still, we came through them. I think that in our response to climate disruption, we see an outstanding example of just the same: a massive exhibit of the human capacity for self-destruction. Let's see if a psychological approach has anything to say about the deniers who insist that climate change is not happening or is not important or is not human made—or that it's now too late to do anything about it.

## Finding Answers in Psychology

So the paradox, again: If the facts are so compelling, why do they not really register? Some would argue that the broad public denial and indifference are caused by the denialists and their vitriolic, oil-funded anti-climate campaigns. That is certainly the case to some extent. There has been a strong and abundant supply of anti-climate messages over more than two decades.[46] But it begs the question: Why are so many easily attracted to contrarian messages? Doubt seems to be an easy sell, and deniers have jumped in to supply contrarian ideas: "It's the sun"; "Climate change is natural—it's happened before"; "It's actually been cooling since 1998" . . . If people responded with scientific rationality based on the massive evidence available at their fingertips, messages like this would die on the vine, no matter how cleverly expressed and communicated.

We need to look closely at the demand side for doubt—the inner reasons why disbelief is attractive. How does denialism—with very few facts, lots of grand rhetoric, and very little scientific brainpower—continue its dark victory?[47]

To find *psychological-level* answers, we can adopt a whole suite of approaches from various psychological schools. The fortress of psychology has surprisingly many rooms, towers, cellars, and appendices. It is not just one unified science, but a pluralistic guesthouse, a chateau even, with kaleidoscopic styles. Some styles of psychology are closely related to literature, story, and interpretation. Others are strictly experimental, empirical, and quantitative. There are rooms dedicated to practical uses. Therapy in particular occupies the main living room, but there are also rooms for self-help groups, and even parapsychology. Some rooms are shared with neurobiologists or artificial intelligence enthusiasts. Other psychologists inhabit large appendices near the philosophical and existential citadels. Still others pair off with economists and social behavior science. In its garden you can also find environmental psychologists and eco-psychologists. Surprisingly, many of these varieties of psychology are relevant to climate change, when you take a closer look.

The remaining chapters in part 1, Thinking, explore four traditions from this diversity that can best help us understand the climate paradoxes: evolutionary psychology, cognitive psychology, social psychology, and the psychology of identity.

# THREE

# The Human Animal, as Seen by Evolutionary Psychology

When I go to my Facebook home page, I get a lot of more or less sleazy ads with photos of women wearing very little. My wife doesn't. The ads Facebook directs to her page are usually for fancy women's clothes. Why? The answer has its roots in evolutionary psychology. Our daily behavior is informed by ancient biology that, as postmodern web marketers have realized, expresses itself freely in clicking patterns, often influenced by gender. They know which imagery to put where to exploit our old but powerful instincts for sex and curiosity. This biology—shaped by our culture—is not only influential in the jungle of web clicking, but strongly present in the way that we relate to real-world dangers—and thus to climate news, too.

The starting point of evolutionary psychology is that natural selection has, over millennia, shaped all living organisms to fit their surroundings. Including us humans. Our ancestors were nomads and hunter-gatherers who for thousands and thousands of years lived in small bands of 20 to 150 people. We can imagine that when game and fruit and other delicacies became exhausted, they simply migrated to the next area. Hopefully to somewhere bountiful, where there was no other hostile tribe. Today, with billions of people, mostly in cities, our prehistoric needs continue to influence our choices and behavior, often unconsciously, even if there are no bountiful free natural spaces left. Since these ancient needs are coded into our genes, say the evolutionary psychologists, they work in the same basic way today even if our modern surroundings and challenges are massively different.

When probing how we consume and behave, evolutionary psychologists highlight five ancestral forces: self-interest, status, social imitation, short-termism, and risk vividness.[1] By understanding how and why this "old mind" continues to shape modern behaviors, more of the climate paradox can become clear. And this understanding can also help shape new strategies about how to resolve the climate paradox.

In 1722 the Dutch explorer Jacob Roggeveen came to Easter Island. At the time the landscape was barren, and the indigenous society had crumbled. Some centuries before, there had been a thriving culture encompassed with lush forests. It had all collapsed. And it was fully human-made. A plausible explanation is that this was due to the big men in the island tribes competing for status among themselves by putting up the now famous monolithic human statues carved out of local rock and moving them long distances. As they got larger and larger, ever more trees were felled to transport the monuments. When Roggeveen landed, "the island was almost completely deforested."[2] But as the forest went away, so did the soils and the habitability of the islands. Unlike the romantic cliché, not all indigenous, or first, peoples have lived in balance with their natural surroundings.[3]

This story, told so well and in great detail by Jared Diamond[4] and others, illustrates that the evolutionary forces in the human animal, if unchecked by culture and good governance, can wreak collective havoc: The combined human tendencies toward self-interest, status, and imitation of others may turn self-destructive. Particularly when you add short-term bias and the propensity to disregard long-term future threats. These evolutionary forces worked well together for thousands of years when the human influence on ecosystems was small, and there were new areas to go to when one was getting exhausted. As the impact of humans via sheer numbers, multiplied by technology and affluence, grows too heavy on a global scale, these forces will push us collectively toward self-destruction if unchecked by culture.

This view may seem to imply that genetic forces will continue to drive humans farther and farther down the road to destroying nature. If so, the earth is like a giant Easter Island and nothing can be done, since human nature is hardwired into our genes. Humans are incapable of caring for the more-than-human, and rarely for more than our own little in-group. Our competitive old mind will drive us to collapse, to a destroyed world, maybe one that recalls Cormac McCarthy's novel *The Road*,[5] until mere, meager survival of the self and offspring becomes the only rule.

But knowing that these evolutionary forces influence us is not enough to predict that this is inevitable in the future. The purpose of outlining them here is the exact opposite: first to explain why it is difficult for humans to respond to climate change; second to find practical, cultural solutions that shape and go with the flow of our evolutionary tendencies. Our problem is not biological, but cultural in character. It may be smarter to understand more of the relevant psycho-biological processes before attempting to modify human action.[6]

## Me: Self-Interest

Let's look first at *self-interest*. From an evolutionary perspective, self-interest is not just egoism, in the sense of "me, me, me," but rather a drive to spread one's genes through sex and offspring. Although people vary in the extent to which they exploit others, most are disposed to make short-term selfish choices in social dilemmas, especially with strangers as opponents. The out-group, the others, are easily seen as *not-me* or *not-us* if we let the old mind run rampant. But although our genes are selfish at the ultimate level, they can and do build organisms and cultures that are capable of collaborating in ways that are kind, charitable, and sustainable.[7] Both wolves and humans traditionally share the food from the hunt generously with their kin, the tribe: I may lose some of my bounty now, but next time around the gesture will be reciprocated, which is good for me, too. Culture may shape the biological short-term self-interest to serve group and community for the longer term.

These insights from evolutionary psychology may help us understand why climate change isn't widely perceived as relevant yet: It's not hitting me, my family, or my pack or group (yet). One plausible reason why only 15 percent of Americans are "very worried" about global warming is that few think they will be personally harmed by it. Of those Americans who are actually "very worried," about eight in ten (78 percent) think it will cause harm to them personally, a very high correlation.[8]

Climate change is perceived as happening elsewhere: the Maldives, Philippines, Arctic, or Antarctic, or in New Orleans or upper Himalayan valleys or the Bangladesh delta. It's about "them," not "us." However, if we could shape and extend the feeling of "us," or possibly utilize the altruistic reciprocity that is linked with self-interest, then we could get on the track of new solutions. Maybe evolutionary self-interest can be harnessed for the common good of climate, too? The second part of the book will investigate this possibility when looking at how to use the power of social networks and stories to extend our sense of the in-group.

## The Flock: Status

*Flock status* is hugely important to humans, just as it is to other social animals like wolves, apes, horses, moose, and large forest grouses like the capercaillie. While writing this, I spent some nights sleeping out in the pine forests at a carefully selected spot, in a sleeping bag under the open sky, to watch the spring capercaillie display just before dawn. The male birds are given to ecstatic dancing, impressive colors on their tail feathers, floating elegantly around, heads raised proudly, praising the sky and the towering trees at this unique place and the spreading dawn light with their rhythmic, guttural clicking song. The dominant male is particularly into extending his presence to fill the whole playground so that all female attention is drawn to "me, me, me." Other males ought to keep their distance. If another male comes too close, loud clashes are inevitable.

Humans aren't that different. But rather than simply dancing proudly with what we've got, we moderns *buy* feathers and colors and dresses and cars and boats and large houses and countless other things to display how big our own playground is. On Easter Island it was about bigger and bigger statues. In our communities, it is about the large home and flashy cars. The challenge—of course—is that status is *relative*. It is not just about having shiny feathers, a large statue, beautiful houses, and fast cars; it is also about having *more* than others.

So the second barrier identified by evolutionary psychology is that winning in relative status *feels* more urgent in the human brain than the long-term threats of climate change. Status, however, has also in some cultures been gained by giving away more than others. Sometimes this is called competitive altruism.[9] Maybe we could shape the ancestral force of status to also benefit nature and one another. How can we have people compete for "green status"? This idea will also be revisited in part 2, when we move on to solutions.

## Do as I Do: Imitation

When it comes to human behavior, *imitation* is paramount. Little sister does as big sister, who does as Mom or Auntie does. Big boys imitate the even bigger boys who imitate the yet bigger boys. The psychologist Solomon

Asch got very famous in the early 1960s for his conformity experiments in which he demonstrated how far most people would go to suppress their own opinion in a group whose members share a different view. When you've seen and heard eight people you respect express their opinion unanimously, all saying that the upper line on the display is longer than the lower, then it is very hard for most of us *not* to imitate them when asked for the ninth opinion. Most will imitate the majority even when our own eyes clearly tell us that the lower line is longer and that those in the majority are all blatantly wrong. Such imitation, of course, isn't news. To express his dislike for it, Henrik Ibsen wrote in *An Enemy of the People*: "The majority never has right on its side."[10] He has been followed by a host of writers who have spilled a lot of angry ink over people's unthinking flock behavior.

Evolutionary psychology highlights that imitating others is efficient. Among social animals, following the majority is good for learning and survival. Birds, bison, elk, ants, and fish do it. Stock investors do it. We all do it. When masses of us join Facebook, is this the "network effect" or "herd behavior"? Because of this deep mimicry instinct, appeals to consumers that they should behave more environmentally often fail. This is because people cannot see—nor feel convinced—that many others are doing it.

Sometimes campaigns or the news brings us messages such as: "83 percent of people are not recycling!" or "Three hundred million plastic bottles are discarded every day."[11] These messages may be well intended, but from an imitation point of view they have the opposite effect: If so many of us don't, then I won't bother, either. Research shows that people throw more litter on the ground in areas that are already littered.[12] Imitation is similarly a barrier to climate action: There is no majority already behaving responsibly that I can imitate. "Everybody else is littering the sky, so I will, too." But imitation isn't fate. We can choose differently. So maybe there are smart ways to start harnessing the evolutionary force of imitation *for* climate action, rather than the opposite.

Many climate deniers claim that the scientific climate consensus is a typical case of imitation and groupthink. They display, however, a very limited knowledge of these psychological concepts,[13] which are commonly but incorrectly used as a label against the climate consensus.[14] Outspoken deniers consistently underestimate the level of scientific skepticism and the diversity of ideas and alternatives discussed in the loosely knit international climate scientist population.

## A Bird in the Hand: Short-Term Thinking

What is short-term thinking? "I want satisfaction now." "A bird in the hand is worth two in the bush." Or in a more original quote from 1546, "Better one byrde in hande than ten in the wood."[15] That humans focus on the short-term isn't news. We were hunter-gatherers long before we were farmers, and while farmers need to wait several months for harvest, hunter-gatherers' labor is often rewarded the same day. So natural selection originally shaped our psychobiology to maximize benefits here and now. But in some occupations, such as modern city and infrastructure planning, people may have to work with a ten- to fifty-year horizon before seeing the physical outcomes of their plans. Or even much longer if you're a climate scientist! Not very gratifying to the human psyche. Behavioral psychology says that the optimal time interval for learning between a stimulus and response is on the order of one to two seconds. That's pretty far from the fifty- to five-hundred-year horizon (and more) relevant for climate change.

Although there are large individual and cultural differences in the ability to delay gratification, people in modern societies still overwhelmingly weigh present outcomes as much more important than distant ones.[16] This leads to underestimating the probability and severity of long-term climate outcomes, which again translates to less willingness to support funding for climate change responses now. In chapter 1, we saw that a majority of respondents say that climate change is a concern, yet give it very low overall policy priority.[17] This may to a large extent be explained by the evolutionary short-term tendency: Many citizens don't seem to mind addressing the economic cost of climate change, as long as it doesn't come out of their own pockets now.

Thus, short-termism is clearly a fundamental barrier to climate change responses. Since climate disruption is—by human standards—a very long-term issue, it falls far down on our list of priorities. That is, until your own house (almost) burns down, as Californians in increasingly wildfire-prone areas are experiencing. The violent crackling is what global warming sounds like, and the odor of the noxious smoke is what it smells like, when you're close to it. Or when millions of people start running out of water due to drought and sinking water tables. Or when a Nepalese valley you live in is so flooded by torrential rains that your fields and all their crops are swept into the river.[18] It's hard to imagine how to harness the universal short-term tendency for the common good. But I believe—along with evolutionary and cognitive psychologists—that it *can* be done.

## Responding to the Spectacular: Our Perception of Risk

When it comes to *evolutionary risk perception*, the evidence is clear that people tend to disregard problems they cannot see or feel. We tend to over-value immediate and spectacular threats while ignoring more impalpable concerns. Poisonous snakes, growling lions, fast-moving things coming toward us, closed-in and narrow places, angry faces, raging fires, sudden loud sounds, a large truck swerving toward our tiny car: All these situations fuel an adrenaline rush. Fight, flight, or freeze? Our entire body is involved, alert, hypervigilant.

Psychologist Daniel Gilbert puts it well:

> The brain is a beautifully engineered get-out-of-the-way machine that constantly scans the environment for things out of whose way it should get right now. That's what brains did for several hundred million years—and then, just a few million years ago, the mamma-lian brain learned a new trick: to predict the timing and location of dangers before they actually happened. Our ability to duck that which is not yet coming is one of the brain's most stunning innova-tions, and we wouldn't have dental floss or 401(k) plans without it.[19]

But this new innovation is still in the early stages of development. The original app that allows us to respond to speeding balls and ominous villains with black mustaches is ancient and reliable, but the add-on utility that allows us to respond to storms, fires, and floods that loom in an unseen future is still in beta testing. Climate communication professionals need to learn how to work with the fear-and-risk neuronal system constructively.

To play on Gilbert's metaphor: Can we move from beta to alpha testing of this risk perception brainware, and get to a working version 1.0 in time to have an effect on reducing climate disruption? The odds are not that bad, I'd say. At least on a social or cultural level. Risk perception is an area that evolutionary psychology has in common with cognitive psychology. Cognitive experiments have exposed a lot of the ways that the ancestral risk system responds to different types of modern challenges. The possibility of reprogramming the genes of individuals seems rather remote—but it may still be possible to "hack" them. Cognitive and neurobiological psychologists love to use computer metaphors for the brain.

## But Genes Are Not Destiny

The human animal comes with the powerful neural wiring of preferring self-interest (me and us!), status (better than thou! more sex than thou!), imitation (herd!), short-term (now! never mind the future!), and spectacular risk (the vivid, not the impalpable!). Climate communicators ignore these exclamation marks from our old mind at the expense of their crucial message. Each represents a powerful barrier, and taken together they contribute a lot of the answer at the individual level as to why more isn't happening. People are stuck in their old mind patterns, and therefore do not prioritize long-term climate considerations. Thus, there is no reason to be surprised or disappointed if consumers continue to prioritize larger cars over more efficient ones, purchase what their neighbors have rather than the greenest product, or do not want to pay small fees now for a long-term future gain. This is how our brains are organized—from the get-go.

Many of our ancestors and their centuries-old cultures developed much more wisdom than these five forces can capture—and stories and rituals to shape that wisdom. But these forces are still common denominators in our makeup, and together they provide a more realistic model of human emotions than the rational information-processing model that conventional climate communication targets.

Genes are all about what worked well in the past, but their current expression is now shaped by language, technology, and culture. They are not destiny. Why not utilize this evolutionary "problem" as a resource for finding future, deeply motivating solutions? Why not create conditions in which we can tap into and work *with* these deeper tendencies in us humans? Rather than trying to correct and discipline them, trying to control and enlighten our animal body? To see how that can be done, or how to hack it, we need to move beyond the ancestral features of our minds and consider the *cognitive* features—in other words, how we perceive, think about, and frame our ideas on issues like climate.

# FOUR

# How Climate Facts and Risks Are Perceived: Cognitive Psychology

How we humans perceive facts does not simply follow from the quality of the science behind them. *Cognition* refers to how we actually think and how the brain processes information. Cognitive psychologists explore, among other things, how we think when we judge available information to make opinions and decisions. That psychology can and should play a role in the climate change debate is not new. It was first pointed out more than thirty years ago.[1] Yet its potential for improving climate communications has still not been tapped.

Those who thoroughly read or hear the latest scientific findings about the increasing pace of climate disruption with an open attitude are very likely to conclude along with many leading scientists that this is increasingly urgent. In 2006, a *Time* magazine cover, showing an image of a polar bear floating on a small piece of ice, issued this alarm to everyone on the planet: "Be worried. Be very worried." Well-meaning readers, newly convinced or with a heightened concern, may have tried to pass on the newfound facts in the magazine's coverage to their network to share their sense of urgency. But this fact-based approach does not easily re-create that same feeling of urgency from one person to the next. Many converts get frustrated that the others just don't get it. Their habitual approach is to then deluge the listeners or readers with ever more facts, statistics, figures, and ominous projections.

What do climate facts look like? Here are some typical ones from NOAA's annual report of 2012 on their website:[2]

- The year 2013 ties with 2003 as the fourth warmest year globally since records began in 1880. The annual global combined land and ocean

surface temperature was 0.62°C (1.12°F) above the 20th century average of 13.9°C (57.0°F). This marks the 37th consecutive year (since
1976) that the yearly global temperature was above average. Currently,
the warmest year on record is 2010, which was 0.66°C (1.19°F) above
average. Including 2013, 9 of the 10 warmest years in the 134-year
period of record have occurred in the 21st century. Only one year
during the 20th century—1998—was warmer than 2013.

- Separately, the 2013 global average land surface temperature was
0.99°C (1.78°F) above the 20th century average of 8.5°C (47.3°F), the
fourth highest annual value on record.

How do you respond to such climate facts after reading some of them?
With soaring engagement? Here is another typical climate fact graph:

What does your cognitive system get out of all this? An impulse to act?
Or fear? An urge to change your life? Or a big yawn?

Not many are immediately able to see the beauty of a breathing earth
by interpreting the tiny cycles in figure 4.1. It shows that during spring
and summer in the Northern Hemisphere there is a kind of inhale of $CO_2$
(the downward half of the jagged line cycle) when the green plants frolic
on carbon dioxide, when sun and water are abundant. During autumn and
winter more $CO_2$ is released as the fungi and bacteria frolic on the organic
remains of leaves and plants drop their photosynthesis levels (the cyclical
rise in the jagged line).

Even fewer would recognize that the four-hundred-parts-per-mil-
lion level that the curve is approaching hasn't ever been crossed in all
of the eight hundred thousand years for which we have clear data. Some
scientists speculate that we might need to go back five or even fifteen
million years to see $CO_2$ at today's levels. On a planetary scale, this is
absolutely breaking-headline news. We crossed the four hundred level
for the first time in May 2013. After 2014 monthly averages are soon
permanently above it.

Raw data, on its own, does not easily give meaning. So how does our
thinking mind construct meaning from dry numbers or ominous curves?
How to connect the dots (or curves) into changed mental models or attitudes with a consistent readiness for action? These are the questions that
cognitive psychology could help us understand from an information-processing point of view.

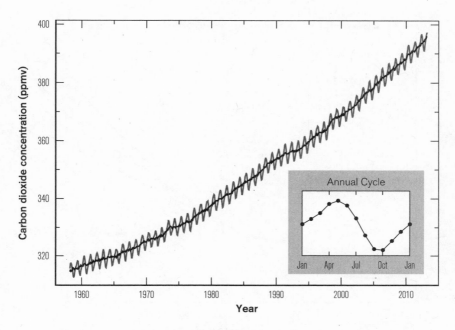

**Figure 4.1.** Atmospheric carbon dioxide measured at Mauna Loa, Hawaii. This is the so-called Keeling curve, after Dr. Ralph Keeling, who started the measurements in the year 1957. It shows the dramatic rise of $CO_2$ levels in the atmosphere over the past five decades.

Figure 4.2 illustrates a typical climate science graph. This graph is intended to show how actual emissions up to 2012 are in the upper range of what was thought plausible back in 2000 (bad news)—and this is despite the dip in economic growth after the financial crisis of 2008–09 (even worse news). The story hidden inside this figure is that we are on our way to the scary, extreme future that IPCC scientists called A1FI, not the good one they called B2. These scenario names (like A1FI, A1T, and B2) are, however, internal to the researcher network, and the stories they tell are utterly incomprehensible to outsiders. Still, graphs such as these are frequently used, since the climate communicators feel they are scientific enough to be defensible. They feel rational and safe when they can present their findings in this way to the public. They may have abundant scientific substance, but they are unfortunately completely lacking in broad appeal.

When the fifth IPCC report was published in 2014, the same communication strategy was repeated; see figure 4.3. But by then the scenario names

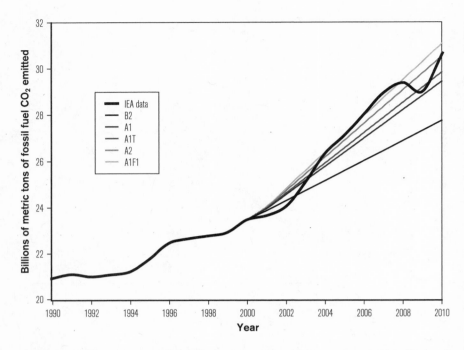

**Figure 4.2.** This graph charts observed $CO_2$ emissions from fossil fuel burning, as recorded by the International Energy Agency (IEA), and compares it with the various scenarios presented by the IPCC in a 2000 report. These IPCC scenarios—named B2, A1, A1T, A2, and A1FI—were created to show a range of possible future climate impacts, from mild to severe. Source: Dana Nuccitelli, http://www.skepticalscience.com/graphics.php.

had changed to RCP2.6 and RCP8.5, as if these new codes would help in getting the facts and uncertainties across. It is hard for nearly all audiences to get the concrete story from such graphs. A lot of climate jargon—like 2C-target, mitigation, or decarbonization—is confusing and off-putting. If your goal is to enhance public understanding and concern, starting with codes and graphs like this is like shooting yourself in the foot before you begin walking.

Just presenting such facts and figures about global warming over and over again has so far *not sufficiently* convinced the general public, journalists, or policy makers about the scale of the problem to create the sense of urgency needed for the required actions. From the cognitive perspective, then, the psychological paradox is not surprising. Selling the science is a mightily difficult task, since the issue of climate change is conveyed as

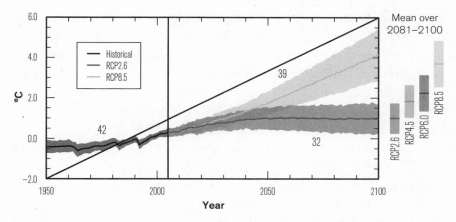

**Figure 4.3.** Global average surface temperature change. This graph depicts emission scenarios and global surface temperatures to 2100. Source: Adapted from IPCC WGI AR5 SPM.7a.

remote, abstract, vague. It lacks salience. You might even call it ghostly or diabolical since it seems to defy nearly all the evolutionary and cognitive hooks our brains use to generate a sense of urgency.

From a cognitive psychology point of view, the main reason why so little is being done and so few individuals take to the streets to protest is that the whole issue is presented as very *distant* to us. This gives a low felt sense of urgency in our information-processing system. This ties into how we perceive climate *risk* perception and how we respond to various kinds of climate messaging, or *framing*.

## The Trouble with Distance

While in a taxi in Cape Town, Janet Swim, a psychology professor who headed the climate task force of American Psychology Association,[3] asked the driver about climate change. Always a good thing to do if you want to understand how people think. "I don't think there's climate change," the driver said. "If there was climate change and sea levels were rising, I would have seen it." Yep, that's how our everyday mind thinks: "People experience weather on a day-to-day basis, and that's how they think about climate change," says Swim.[4] This little story is confirmed by systematic research: It's not just taxi drivers. The polls on "belief in" or "worry about" climate change are highly influenced by temperature variations over the previous three to twelve months. In addition, the number of editorial and opinion articles that agree with the expert consensus on climate change is also found to be correlated to temperatures. When the weather is unusually hot, people get concerned about global warming. In cold spells, concern wanes.[5]

Therefore, the slow-moving, decadal climate change—as distinct from weather—appears distant to most people in a number of ways.[6]

It feels distant in *time*: The years given—2050 and beyond—seem very far away. It is distant in *space*: Typically the effects are strongest in the Arctic, Antarctica, the Pacific Ocean, or in other locales far removed from where most people live. It is *invisible*: $CO_2$ is colorless and odorless. It is neither caustic nor poisonous. The gases are very *rare*, only around four parts $CO_2$ per ten thousand in the atmosphere. It is depicted in very academic *abstract* terms: measured in so-called ppms (parts per million). It works through more invisible radiative forcing in the atmosphere, measured in $W/m^2$ (watts per square meter). Whatever this all means, it cannot be seen, felt, or touched.

The combined outcome makes us feel *helpless* since—even if we stopped emitting now, they say—its delayed effects (including our grandfathers' coal burning from last century) will continue to trouble us in decades and even centuries ahead. This helpless feeling grows from the recognition that the climate disruptions are very distant from our own *locus of control*. Finally, nearly everyone (in wealthy countries) is implicated, and all have to change for the shared results to benefit all. It is the big guys'—congressional representatives' or international leaders'—responsibility to do something about it. I don't know them myself, nor do I know someone who knows them. Thus it is also *socially distant*. It's not about myself and my friends.

If we wanted to create a threat too ghastly and ghostly for our perception and thinking to handle, it is hard to imagine designing something better than the presence of too much human-caused $CO_2$ into the air.[7] Try to hit it and we're just hitting empty air. It's there, but it seems utterly intangible, unreal and real at the same time. Distant in time and space, invisible to the senses, abstract, and socially far from our influence.

Psychologist and economist Daniel Kahneman has summarized a lot of cognitive research into describing our two minds, or our two main cognitive systems: the *fast* and the *slow*.[8] The fast system is a kind of quick-and-dirty thinking based on rules of thumb, habit, gut reactions, and biases.

The links to the "old mind" of the evolutionary psychologists are many. It is mostly unconscious and makes snap, intuitive judgments inferred from our past experiences and emotions. If there is a right answer, we are as likely to be wrong as right when we use this fast system. The slow system, on the other hand, is more rational, linear, logical, and cumbersome. If there is a right answer to a rational question, the chance for getting at it is much better with the more conscious slow-mind system. When the two can work together, they provide us with a broader view of the world around us.

With regard to the climate problem, we really have to do a lot of slow, deliberative reading, noticing, and thinking to get a sense of urgency. There is very little in the slow global warming itself that raises red flags, evoking strong visceral reactions and rapid responses from our fast system, unless we happen to be in the grip of a typhoon or hurricane.[9] Very few are, and only for short durations. Unfortunately, the climate disruptions are not caused by a tyrant with a plan to destroy the world through toxic emissions. If they were, then we could have declared him our global enemy number one and bombed him out of existence. Neither are they caused by gay sex, terrorists, drug barons, or other popular targets for raising political fury.

Lengthy scientific explanations of why the local weather can be unusually cool for some weeks, even while we're facing global warming, seem to be of little avail. A Columbia University study looked closely at how individuals make up their minds with regard to climate change. Do they rely on recent weather or on scientific climate change analysis? It was found that most people clearly rely on the recent weather.[10] After periods of unusual warmth, people tend to be more convinced and more concerned about global warming. Conversely, after unusually cold periods, people's views tend to go in the opposite direction. Despite the clear climate facts, the fast system's mental shortcuts win out.[11]

New Yorkers' interest in climate surged after Superstorm Sandy in 2012, only to fall back to normal during 2013.[12] For farmers in Norway, interest in climate change surged after an autumn in which repeated flooding made it impossible for them to get their tractors out onto the fields for months on end. This brought climate change to their doorstep.[13] We may then predict that when we again see a year with fewer rainy months and less flooding, so farmers again can drive on their fields, then their overall interest in climate change will recede. Elsewhere, those on drought-stricken farms are likely to become less focused on climate when the rains finally arrive.

Most people tend to avoid *slow*, complex thinking by using a simpler and *fast* evaluation, to use Kahneman's metaphors.

In other words, if it's in my face, or an inescapable threat, I'll be concerned about it. Anything that makes a threat appear distant from me lowers the felt concern. One global survey[14] found that people living in coastal countries (read: countries that are surrounded by water like Indonesia, Bangladesh, Thailand, Philippines, and the Maldives) are way more concerned about global warming than those who aren't. In other words, if you live in America's heartland, you're not too concerned about the ocean levels rising—unless you have relatives on the Florida coast. If you're living on a flat island or low-lying coast, then sea surges are an immediate threat to your existence. So you're way more motivated to buy a product manufactured with reduced emissions, or that limits your own use of fossil fuels. "In middle-America, not so much . . . Folks in countries with water shortages are worried about, um, water shortages. Folks in countries with much water pollution? Right, worried about water pollution," comments analyst Patrick Hunt.[15]

In order to communicate climate urgency, facts and figures must, at the very least, be tailored to fit with the way in which different groups of people process information. Communication must be designed to fit how we cognitively perceive risk, and form attitudes and opinions. This is difficult, but far from impossible. Climate scientists could partner with psychologists, sociologists, artists, and social scientists to communicate the science through visualization, frames, and stories that foster action and hope rather than despair and denial.[16]

For instance, when discussing $CO_2$ levels, it has become customary among people with climate knowledge to refer to ppm levels. Not many know that $CO_2$ levels in the atmosphere were less than 270 ppm before the industrial revolution. Now we're past 400 ppm. But this designation is abstract and thus cognitively distant: 270 or 400 of something invisible— who cares? A better way of communicating it could be to say that we now have over four parts per ten thousand, while it used to be well below three. That's close to a 50 percent increase over the last century and a half. Or even more visually: Let's imagine that $CO_2$ is a brown haze heavier than air. All $CO_2$ thus falls down to earth and creates a blanket layer, a pool around us that we are "swimming" in. How high would this layer be if spread out evenly all around the world? It used to be around fifteen feet before the industrial revolution. But humans have now added around seven extra feet so we're today immersed in a dense twenty-two-foot-high fog all over the world. Feels a bit more suffocating than "400 ppm," doesn't it?

## Risk Is a Feeling, Not a Number

Climate communication has struggled to get the real risk across. Somehow, the biggest challenge of our time doesn't register as a real risk in people's mind. Yet how we actually perceive risk has gotten a lot of attention from cognitive psychologists.

One key finding is that *people exaggerate spectacular but rare risks and downplay common ones*. In 2009 a very few Toyota owners started to report that their cars were behaving abnormally: Suddenly the car would seem to accelerate strongly out of control. A crash near San Diego in August that year killed a California Highway Patrol officer and three family members. Sudden acceleration incidents involving Toyota cars and trucks were claimed to have taken from nineteen to twenty-one lives since the 2002 model year, the *Los Angeles Times* reported.[17] All the media then roared at the same time. Toyota was forced to recall nine million cars produced between 2004 and 2009. Sales of new Toyotas plummeted. Prices of used cars went off a cliff. The recurring news created mounting fears that my seemingly friendly car would run away with me and kill me. It tarnished Toyota's reputation for leading quality. But exactly how dangerous was the issue?

Let's assume a worst-case scenario: that sudden unintended acceleration was the cause for all twenty-one deaths in the period with nine million cars recalled over seven years. This translates into a probability of dying from sudden acceleration of 1 in 3 million per year if your Toyota is not recalled and modified. How dangerous are odds of 1 in 3 million? Well, the chance of dying in your treacherous bathtub in any given year is about 1 in 685,000; in a plane crash, 1 in 354,000; from falling down stairs, 1 in 150,000; from choking while eating, around 1 in 100,000. The likelihood that you will die in any ordinary car crash unrelated to unintended acceleration is 1 in 22,000 in any year in Norway, and around 1 in 6,500 in the United States.[18] But thanks to the spectacular acceleration stories (all involving images of high-speed, out-of-control car crashes), driving your new and relatively safe Toyota (1 in 3 million) suddenly *felt* like driving a coffin. It could kill you at any moment. The malfunction was a serious issue, of course, but at the same time it is much more dangerous to walk down the stairs at home or enter the traffic in any type of car at all. So this was a case of our public risk perception getting the proportions all wrong.

In the case of future climate disruption, we also get the proportions all wrong, but unfortunately in the opposite direction. Climate science is telling us that if societies continue emitting as they do today, there is a 95 percent probability that we will see much graver consequences during this century.[19] That's a close to 1:1 level, versus the 1:3 million level for Toyota trouble. But psychologically, risks are feelings, not numbers.[20]

*Personified risks* are perceived to be greater than abstract or anonymous risks. Learning that the risk of dying in a car accident is 1 in 6,000 or 1 in 20,000 is not very provocative. If, however, one of your friends dies in a car crash, then suddenly it is a personal tragedy. It is no longer a distant probability, but very concrete and emotional. Also, near-death events on the roads, in which you or someone you know barely survives, leave a strong mark. Some never want to sit in a car again. Or do not want to get behind the wheel for years.

Further, people underestimate risks they willingly take and overestimate risks in situations they can't *control*. Many feel much safer and more comfortable when they drive themselves, rather than being a passenger.

Most people are more afraid of risks that are *new* than those they've lived with for a while. In the summer of 1999, New Yorkers were extremely afraid of West Nile virus, a mosquito-borne infection that had never been seen in the United States. By the summer of 2001, though the virus continued to show up and make a few people sick, the fear had abated. The risk was still there, but New Yorkers had lived with it for a while. Their familiarity with it helped them see it differently. Similarly, news about climate change has been around for some decades now. It is not a novel scare any longer. It hasn't killed us yet. Habituation has set in.

Most, too, are less afraid of risks that are natural than those that are artificial and human-made. Many people are more afraid of radiation from nuclear waste, or cell phones, than they are of radiation from the sun, a far greater health risk. Many are more worried about extremists and terrorism than climate. To try to calculate the odds of being killed or injured by terrorists, Michael Rothschild, a former business professor at the University of Wisconsin, worked out a couple of plausible scenarios. For example, he figured that if terrorists were to destroy entirely one of America's forty thousand shopping malls at random *per week*, your chances of being hit (if you are a US citizen) would be about 1 in 1 million or higher. Rothschild also estimated that if terrorists hijacked and crashed one of America's eighteen thousand commercial flights *per week*, your chance of being on that crashed plane would be 1 in 135,000 per year. The risk of dying from heart attack or heart failure is more on the order of 1 in 400 per year.[21]

Most people are more afraid of a risk *imposed* on them than a risk *they freely choose to take*. Smokers are less afraid of hazards from their own smoking than they are of asbestos and other indoor air pollution in their workplace, which is something over which they have little choice.[22]

Last, people overestimate risks that are being much talked about and remain in the public spotlight. Each time there is an outbreak of infectious disease, such as Ebola, SARS virus, mad cow disease, or avian flu, media attention generates a lot of public anxiety. The risks are perceived by most as high, even when the risk of personal exposure is next to zero and minuscule compared with that of getting into a car. Some years later—even if the biological risks are still there, or even higher due to growing urbanization and globalization—the felt risk has evaporated. Fears, danger, and risks are thus created by people and the media in how we talk to one another. We rarely relate directly to a future quantitative risk, but always to images and stories of danger. Risk is a feeling.

In sum, when it comes to climate risk perception: People are prone to exaggerate risks that are spectacular, new and unfamiliar, personified, beyond personal control, much discussed, immediate, and sudden as well as those that affect them personally and are imposed by a clear enemy. They tend to downplay risks that are dull, common and familiar, anonymous, somewhat controllable, not much discussed, long-term, gradual, and natural, as well as those that affect others and lack any clear bad guy.[23]

What makes the climate crisis ghostly is that it seems to press nearly all the wrong buttons. It is about abstract, imperceptible, and gradual changes in weather trends from decade to decade. It is anonymous and not personified. It is beyond anyone's control and reach. It is rarely talked about at social events at the in-group level. It has a complex indirect impact on primarily far-off strangers, not us and our group. It is old and yesterday's news.

(Public concerns about climate change were actually higher in 1989, a year after scientist James Hansen, then with NASA, testified before Congress that climate change had indeed begun, prompting headline news.) Finally, there is no real enemy. If there is an enemy, it is none other than ourselves. All moderns participate in economies that are dependent on energy from fossil coal, oil, or gas. But declaring ourselves the enemy will hardly make a killing . . .

No wonder, then, that the climate science message—even if repeated in the media after each extreme event—doesn't really stick in the mind as a high concern with a stable sense of urgency. Our cognitive risk system will—by itself—shake it off as distant, and focus on something nearer, personal, and spectacular. *Perception* of risk thus is often very different from objective, or calculated risk.[24]

Most people relate to climate risks through our fast system, rather than through our slow but thorough system of thinking. This was pointed out long ago by the Greek philosopher Epictetus: "Men are disturbed not by things, but by the view which they take of them." For communicators of climate risks, it is not sufficient just to be aware of such cognitive biases. The larger issue is how to leverage the biases in order to strengthen the climate change message by compensating for the inherent weakness of the climate issue itself. Risk psychology can contribute to new climate communication solutions and ways to redesign the message to create a more sustained sense of urgency (see chapter 10, Reframing the Climate Messages).

## Framing: Don't Think of an Elephant

Is the glass half full? Or is it half empty? Or maybe just right? We can view the beer glass through each of these three frames. The volume of beer is the same, but the meaning given to it depends on the frame.

Framing is everywhere in human communications. News about "mad cow disease" elicits greater fear than reports about incidences of bovine spongiform encephalitis (BSE) or Creutzfeldt-Jakob disease.[25] The three refer to essentially the same sickness, but the first has a stronger framing effect than the two latter. Another example: people are more positive if some-one is described as a "paperless worker" rather than "an illegal immigrant."

One of the best exercises to experience the effects of framing is given by the cognitive linguist and framing expert George Lakoff. He sometimes gives his audience a simple instruction: "Don't think of an elephant!"—as he recounts in a book of the same name.[26] Now, if we're in the audience, we're either secretly thinking of an elephant (because we can't resist the urge) or actively trying to follow the instruction by thinking hard about something else—say, a lion or dragon or dog or airplane or bicycle or something. We must concentrate pretty hard on it; otherwise our minds slip back to that stupid elephant, pink or whatever.

The whole point is that being told *not* to think of an elephant makes it nearly impossible to think about anything else. The message envelops us. The social situation has been changed by the words in the instruction. Now this image somehow envelops and dominates the conversation, whether we like it or not.

Many words, like *elephant*, evoke a frame, which can be an image, an idea, or a group of ideas. For example, elephants are large, have floppy ears and a trunk, are associated with savannas and circuses, and so on. The word gets its meaning relative to that frame. Even "when we negate a frame, we evoke the frame," says Lakoff.[27] Framing is inevitable. Our thinking and our conversations usually happen inside one frame or another defined by context. The question in our context, of course, is: How has cognitive framing influenced climate communication and contributed to the paradox? And going forward, what is the best frame to use: global warming or climate change?

**Ringside:** *Global Warming* **or** *Climate Change*? An in-depth study by Yale researchers found that almost without exception, *global warming* is a more engaging frame than *climate change*. Compared with *climate change*, it generates greater certainty that the phenomenon is happening, greater understanding that human activities are the primary driver of warming, a greater sense of personal threat as well as more intense worry, a greater sense that people are being harmed right now by warming, a greater sense of threat to future generations, and finally greater support for both large- and small-scale actions by the United States.[28]

But the frame *global warming* also has its drawbacks. It creates more negative feelings, and then directs our fast-thinking system to look to surface temperatures given by weather.[29] It reinforces the expectation that every year, seasons will be warmer than the previous. But we know that weather is very variable. The earth's energy flows are unfathomably complex over seasons, decades, and centuries. Global average surface temperature is only one out of ten or more main ways to measure the earth's thermal balance.[30] And seven-tenths of the earth's surface is water anyway.

Scientists often prefer the term *climate change* for technical reasons. But this more neutral frame brings with it an implication that climate has always changed in evolutionary times. Frank Luntz, a Republican strategist, wrote a memorandum urging fellow Republicans to exploit the weakness of *climate change*: "While global warming has catastrophic connotations attached to it, climate change sounds a more controllable and less emotional challenge."[31] It leads to thoughts such as: *What's wrong with change? It's natural. Are you against change?* Because this message has been repeated year after year since 2002, it has by 2014 become the most frequently used anti-climate-change myth.[32] Further, focusing on the word *change* can create doubts about agency. Do humans or natural variations cause it?

Communicators should be aware that the two frames generate different interpretations among the general public and specific subgroups: The use of the term *climate change* appears to actually *reduce* issue engagement by Democrats, independents, liberals, and moderates.[33]

Neither term may be suitable for what climate science is revealing. The last time the air had such an astonishingly high concentration of $CO_2$ was well before we *Homo sapiens* roamed the earth.[34] At that time, the world was likely five to eight degrees Fahrenheit warmer with sea levels 15 to 130 feet higher. That's another world, ecologically speaking, than the one that humans know and that our civilization is built for. Rather than either *climate change* or *global warming*, maybe a wholly new term is called for.

The *ozone hole* was a framing that caught on and helped the issue find a successful resolution. This framing helped us see that our everyday chemicals "drilled a hole" in the earth's protective ozone layer, so that dangerous radiation cut through to ram our skin, increasing cancers. Good commercial substitutes for the dangerous chemicals were available, and so the industrial shift to the new and less destructive substances happened rather swiftly. Everybody could easily help by simply choosing CFC-free spray cans. The ozone layer is now recovering, if slowly. A true success story of caring for a clean air, partly thanks to good framing.

The simple point is that we're way outside any conditions humanity has experienced or really understands. During the last hundred years, humanity has rapidly created a climatic break with the past, disrupting the slow and gradual evolutionary changes.[35] *Global burning*, *global weirding*, or *climate disruption* could thus be more appropriate frames. Personally I prefer the last frame, and tend to use *climate disruption* for the above reasons.

**Ringside: *Carbon Tax* or *Earmarked Offset?*** The choice of frames is crucial not just to communicating the science, but also to facilitating dialogues on policy solutions. Take the frame brought up by the term *carbon tax*. It sounds ominous in many people's heads because taxes have long been framed as negative, as excessive costs, as money that the state takes away from you. I've suffered a loss. Lakoff has pointed out the ramifications of the *tax relief* frame used by conservatives.[36]

New studies have indicated that if we frame the carbon tax as *earmarked* for green energy or as *carbon offsets*, attitudes toward it change significantly. *Earmarked offsets* creates a different image in your head than *raising taxes*. People's willingness to pay thus changes depending on the framing.[37]

**Ringside: *Email Theft* or *Climategate?*** Maybe the most dramatic example of the framing effect on climate communication—and a huge contributor to the recent strength of the climate paradox—was the Climategate issue. It broke just a few weeks before the Copenhagen 2009 climate conference. Hackers targeted the Climate Research Unit at University of East Anglia, stole emails, and then posted links onto contrarian blogs. Only a few hours later, a commenter on Wattsupwiththat.com, a leading anti-climate website, started recommending that the new word *Climategate* be used as a strategic framing device:

It's nice that someone has dropped a big comb of honey onto this ants' nest. But all of the inside chatter in these emails, revealing though it may be to those lapping it up, won't mean a thing to the average news reporter, media outlet, and the public in general. What's needed is a panel of unimpeachable individuals (i.e. no one named in this data drop) who can go through the file, vouch for its authenticity, and issue a quick white paper explaining its implications. The media are clueless. They need to be helped to understand

the significance of—CLIMATEGATE! LEAK OF SECRET EMAILS
SHOWS TOP CLIMATE SCIENTISTS ENGAGED IN MASSIVE
FRAUD! GLOBAL WARMING WAS HOAX DESIGNED TO
ENRICH POLITICIANS AND RESEARCHERS!/Mr Lynn[38]

The framing strategy proposed by Mr. Lynn succeeded. The idea
went viral. Over the next several hours, the term *Climategate* propagated
through blogs and as hashtag #climategate on Twitter. It soon pushed out
email theft and *East Anglia* as a referable moniker. Just as when Richard
Nixon said "I am not a crook" and everyone decided he was a crook, now
each time a climate activist or scientist responded to Climategate, everyone knew that climate researchers were implicated in the scandal. All the
climate scientists involved with the IPCC were now unmasked, according
to the gleeful deniers. With the early adoption of such a straightforward
frame around the anonymous email theft, its victims lost credibility by
its very name. It quickly went viral from online blogs to a runaway roller
coaster on paper and TV media. Environmentalists challenging the escalating meme could do little to stop its spread, and in fact may have inadvertently solidified its effect as a framing device by using their opponents'
frame to argue against it.[39]

Climategate as a framing for stolen emails with scientist chatter may
not have had a direct impact on the Copenhagen negotiations, but it did
have a major impact on the perceptions of climate science particularly in
English-speaking countries such as the United States, United Kingdom,
and Australia. No less than three independent investigations into the
emails found no evidence of wrongdoing or unethical conduct among the
scientists. But the combination of the failed Copenhagen conference and
the highly successful framing from the anti-climate activists was seemingly
effective in undermining the public confidence in climate science in the
2010–12 period. News about the clearing of the scientists by the investigations never reached the same level of public interest as that evoked by the
cognitive framing of climate scientists being caught out in a scandal. Who
actually stole the emails and posted them online has never been discovered,
and the police closed the unresolved case in July 2012.[40]

Rather than complaining about denialists using framing effectively
to manufacture doubt on climate urgency, climate communicators may
learn from it how effective framing can be used. And start using framing to

create care for the air, not just serving those who work hard to ruin it and everyone else along with it.

## Backfiring Frames: Losses, Cost, and Sacrifice

Another important frame to address is *climate action is costly*. Let's say you have invested ten thousand dollars in a company's shares. The shares then go up one thousand dollars. Are you happy? Yes, of course. But let's say that the shares drop a thousand dollars. Are you unhappy? You bet! But are you just as happy for the gain of a thousand as you are unhappy about the loss of it? Rationally, you should be. It is the same amount of money. But psychologically, no! We hate losses much more than we enjoy gains.

A number of cognitive researchers, led by Daniel Kahneman,[41] have shown that we are consistently too loss-averse. People care more about losing a dollar than gaining a dollar. About twice as much.[42] The same applies to good bottles of wine, houses, and all kinds of trades: We react more strongly to losing than to acquiring them.

But how has the climate policy been framed for the public? As massively costly. The message that has come from the calculators is that we'd *lose* a lot of money via higher gas prices, and that $CO_2$ capture and storage equipment is very *expensive*. Putting a hefty *tax* on carbon would increase all kinds of *costs*. We'd *lose* economic growth. Cutting emissions would create *job losses*, losses for industry, loss of competitiveness, loss of wealth. A thousand times, economists and politicians have told us that the costs are (too) high. We are richer than ever, but still don't have *enough*. Money is *scarce*. We'll lose too much. The whole apparatus of cost–benefit analyses and integrated assessment models has tried to balance *losses* and gains, to find the most *cost*-effective measures.

And the gains? They are portrayed as far off into an uncertain future way beyond midcentury. Which doesn't really count, either, in economic terms, due to the high discount rates often employed to discount the future values down to net present value. For instance, the future costs of vast climate destruction become very small when adjusted to present-day value.

The estimated damages may be right or wrong, in ecological terms. The discount rate could be high or low in financial terms. But from a psychological framing perspective, what is exceedingly clear is that *the framing of climate policy as losses and costs* has so far won the discussion. When inside this frame, there seem to be no gains at all. And then it doesn't pass the what's-in-it-for-me test. Even those with lower estimates of costs have had to argue within this frame, saying that the policies are *not* costly. These attempts to negate the frame have actually just reinforced it in the mind of the public.

For instance, the Australian government elected in 2013 promised to abolish the country's carbon tax that was introduced a few years earlier. This was based on the election promises by Prime Minister Abbott to the citizens of Australia to "axe the tax," a strong framing. He won popularity on that promise of cutting costs and got elected. He delivered on his word and killed the carbon tax in 2014. From a cognitive point of view it is damn hard to make a message of *increasing the carbon tax* attractive within the obstructive cost-based framing. Thus by pushing the cost and loss frame, many economists have unwittingly contributed to the broad rejection of their favorite remedy: the carbon tax.

What about the environmentalists? The typical frame—also unwitting—is that of impending catastrophe and all the *sacrifices* we must make to avert it. Since climate change is so huge and destructive, they've said, we must do everything we can to stop it. We consume too much, and therefore should eat less meat, fly less, have no or smaller cars, save energy at home, and reuse and recycle rather than buy new stuff. Yep, that's right: No more tender beef for you. You lost. We all have to sacrifice our over-the-top life-styles, get real, live decently and prudently. Eat your broccoli, get on your bike, and smile.

Again, this message may be right or wrong, in ecological terms. But from a framing perspective, what comes across to the public in historically Christian countries is a sermon about sin and repentance. We must sacrifice our sinful pleasures now in order to be purged for a clean and bright but very far-off future. It's the same framing that has been repeated for a thousand years or more in Christian cultures . . . without any noticeable improvements in behavior. We're inoculated against messages inside that framing.

Cognitive psychology tells us that such reactions are what we may expect when our framing is primarily one of loss, cost, and sacrifice—because we hate losses more than we love the corresponding gains.[43] Like it or not, but Western people's minds simply seem to operate this way. We can either accept that and work with it, or try to fight it on moral grounds.

I think the smarter option is to work with it. For instance, when describing climate-friendly actions, messages could exploit the negative force of the cost frame, rather than suffer it. This can be done by focusing more on the potential to cut losses than on the corresponding potential gains. For instance, people might move away from fossil fuels more readily if the message is 'Cut your fuel costs or heating bills,' and if they can immediately see the gains of installing options like insulation, solar water heating, and solar panels.[44] This is now working in the case of marketing solar panels that, with a loan, can be installed at no upfront cost to homeowners. You cut your bills right away.

In other words, when we become aware of how perception, risk, and framing together influence the mind, we can start crafting solutions. New ways of envisioning climate change can bring the message all the way to our own doorstep, feet, and lungs. We can start to discard the framings that maintain current barriers and embrace the new ones that support solutions.

# FIVE

# What Others Are Saying: Social Psychology

An experiment was once conducted to see if even moths had social antennas that influenced their behavior. First the experimenters measured how many trials a lone moth needed to learn how to fly through a labyrinth to some lovely nectar-ish food at the opposite end. Then, with this baseline set, they added a new condition. They placed four other moths inside a cage at the labyrinth's entrance, just beside the lone moth, and ran the test again. All the moths had been able to see, smell, and hear one another and communicate before the lone moth entered the labyrinth. The results? The presence of extra moths "watching" gave a boost to performance: The moths now learned the route to food much more quickly.

Modern cultures may have a strongly individualistic bias, but we humans remain social animals through and through. It's a trait we share with most other life-forms in nature, but are we more social than other beings? Humans have an unusually big brain compared with body weight, and one proposed explanation is that even if our neuronic capacity seems excessive, the complexities of social relations in fact make it necessary.

One of the most famous experiments in organizational psychology led to the identification of the so-called Hawthorne effect. From 1924 to 1932 at The Hawthorne Works, a Western Electric factory outside Chicago, researchers interested in the evolving movement for scientific management wanted to measure if workers could become more productive with higher or lower levels of lighting. But a funny thing happened: The workers' productivity seemed to improve when *any kind of change* in light was made but then slumped when the studies concluded. How could it be that productivity went up both when lighting increased and decreased? And then disappeared afterward?

The explanation that eventually dawned on the researchers, and made the studies famous, was this: It was the extra attention from management and the researchers themselves that gave workers the boost in motivation, much more than any ups or downs in the intensity of light. I would liken this to the lone-moth effect: In the presence of others, behavior, attention, and performance are changed. Social attention is a very powerful motivator.

Ever wondered why it is so difficult to do your daily exercise, change your diet, re-insulate the house, or do yoga meditation on your own, and so much easier if there is a group of people you like to join at an appointed time? Now you know: You're tapping into the deep well of socially motivated behavior.

Social psychology is the study of how people's thoughts, feelings, and behaviors are influenced by others[1]—even if the presence of others is only imagined. Showering or driving alone, I imagine myself singing to a thrilled audience, and so I take care, unconsciously, to hit the right notes. Social psychologists understand this, and if more of them had worked alongside the climate researchers, then maybe the approach to climate science communications would have been very different.

Climate scientists, and the excellent mathematical climate modelers among them, are social animals, too, in their personal lives. But in their scientist role and when communicating their final results, they tend to believe that each single human weighs the evidence rationally on independent scales. When the evidence is clear, they expect people to conclude on this basis alone—like the experts in their own scientific field—that climate disruption is real and urgent enough to agree and take action.

But ordinary folks, for the most part, don't work this way. People trying to digest climate news and evidence from the Intergovernmental Panel for Climate Change—even if they're listening or reading by themselves—have others present in the back of their minds: *What shall I say, tweet, or blog about this to others? What will my colleagues and friends think about this?* The IPCC climate science texts themselves are actually written in a very social, interactive group process. But the writers seem to have very limited reference groups in their heads. They have been thinking: *What will the peer reviewers say? And the critics?* Each sentence is crafted with a narrow academic target audience in the back of the mind.

Okay, you could say, they are purposely delivering cut-and-dried data that the media or others will translate into more human terms. But is that what it means to adequately communicate facts? And even if you answered yes, then what about the IPCC's role in summarizing its findings? With each major report it releases, it also releases a summary for policymakers. Consider this passage from the first page of the IPCC *Fifth Assessment Report, Summary for Policymakers*, released in 2014:

> Confidence in precipitation change averaged over global land areas since 1901 is low prior to 1951 and medium afterwards. Averaged over the mid-latitude land areas of the Northern Hemisphere, precipitation has increased since 1901 (medium confidence before and high confidence after 1951). For other latitudes area-averaged long-term positive or negative trends have low confidence . . .[2]

After having read this, can I tell my buddies that it rains more now than before? And if asked how I know this, can I point to the above paragraph? The language that this is expressed in is very precise, but the manner in which it is communicated does not lend itself to inclusive social interaction.

If this social nature of humans is such a basic, simple, and self-evident fact, why then is there no care for the audience's values, knowledge, or beliefs exhibited in the IPCC reports, and in the summaries in particular? No attempt has been made to address the concerns of ordinary people. The authors instead detach knowledge from meaning.[3] There is no mention of the word *social* anywhere, and *human* is used only to indicate the "human contributions" to or "human influence" on the climate system. The whole report is a collection of universal climate facts written as if humans were fully outside climate (except for causing the $CO_2$ emissions). As if this text were not part of a social discourse. This is reinforced by the only illustration, the front page showing a distant, dirty melting glacier with no people in sight and very far removed from any apparent social context.

As climate scientist Rasmus Benestad (of the science blog *RealClimate*, realclimate.org) observes:

> The authors of the SPM [*Summary for Policymakers*] are experts at writing scientific papers, but that is a different skill to writing for non-scientists. Often, the order of presentation for non-scientists is opposite to the way papers are presented in sciences. A summary

should really start with the most important message, but this SPM starts by discussing uncertainties. It is then difficult for non-scientists to make sense of the report. Are the results reliable or not?[4]

In other words, even if the IPCC language makes sense on a factual level, it doesn't make sense on a social one. Data has been conveyed, but not context. So what does that mean for someone reading the summary (and I hope a few will)? I do not deny the genuine personal curiosity about natural science that motivates so many to read and respond to climate data. But the most important questions swirling in most readers' heads will still be these: *What shall I say about this to others? How will my colleagues and friends react to this?* Facts do not exist in social vacuum, and to create local meaning they have to be communicated with this in mind.

## Climate Attitudes: Alarmed, Concerned, or Dismissive?

What's your attitude to oranges? Maybe you love them, think they taste great. Eat one a day? Start the morning by drinking fresh orange juice? You know they contain vitamin C and therefore are healthy.

Psychologists define attitudes as a learned tendency to value things such as oranges in a certain way. Or, put a little more technically, "Learned predispositions to respond in a favorable or unfavorable manner to a particular person, behavior, belief or thing."[5] Attitudes are often clearly positive or negative. But at times they can be mixed.

Attitudes consist of three main parts, as captured in what psychologists call the ABC-model:

- An **affective** or emotional component: What feeling is connected to the thing, person, issue, or event? (I really do love oranges.)
- A **behavioral** component: What kind of action or readiness for behavior lies dormant in the attitude? (That daily glass of juice.)
- A **cognitive** component: What thoughts, knowledge, and beliefs come up from memory when attending to the issue? (Yep, the vitamin C.)

An attitude is strong and consistent if all three components are aligned: "I love oranges and eat one daily because they contain vitamin C."

Attitudes can also be explicit and implicit. Explicit attitudes are those that we are consciously aware of and that clearly influence our behaviors and beliefs. Implicit attitudes may be unconscious, but still have an effect on our beliefs and behaviors; they influence which experts or facts we pay attention to. An attitude can be learned or adopted fast, but—once learned—changing it on our own can be pretty cumbersome and slow, as Kahneman would phrase it.

So what's your attitude to climate change? And how did you get it? Many people feel concerned; this has been measured in many different studies in different countries, even if there are no global studies that use the same wording and methodology over time.[6] We know that globally a majority of people, 54 percent, see it as a major threat to their country (see figure 1.3). The results are more mixed when people are asked whether the climate changes are human-caused. And the extent to which people feel guilty is yet more difficult to measure.

A team at Yale University has looked extensively into social attitudes on climate change in America. Anthony Leiserowitz and colleagues used a set of questions to divide American adults into six groups:[7]

1. **The Alarmed.** Thirteen percent say they are well-informed about global warming (cognition). They are convinced it is happening, human-caused, and feel it is a serious and urgent threat (affect). The alarmed try to do something in their own lives and support an aggressive national response (behavior). Here all three components of the attitude are largely consistent.

2. **The Concerned.** Thirty-one percent are also convinced that global warming is a serious problem and are moderately well informed about the issue. But while they support a vigorous national policy, they are distinctly less involved in the issue—and less likely to be taking personal action (weaker behavior component)—than the Alarmed.

3. **The Cautious.** Twenty-three percent tend to think that global warming is a problem, but are not certain. They have given some thought to the issue, but not extensively. They don't view it as a personal threat, and don't feel a sense of urgency to do much about it. Other concerns, such as the economy or security, get higher priority.

4. **The Disengaged.** Seven percent say they haven't thought much about the issue at all, and don't really know much about it. It hasn't really

registered on their radar other than as noise. They believe that the consequences are decades away, and don't see why they should bother. But the disengaged are also the most likely to say that they could easily change their minds about global warming.

5. **The Doubtful.** Thirteen percent are evenly split among those who think global warming is happening, those who think it isn't, and those who don't know. Many within this group believe that if global warming is happening, it has natural causes. They believe global warming won't harm people for many decades into the future, if ever, and say that the government is already doing enough to respond to the threat.

6. **The Dismissive.** Thirteen percent are—unlike the Alarmed—sure that global warming is not happening. They feel negatively about the issue. They are firmly convinced that warming is not a threat to either people or non-human nature, and are strongly opposed to any climate mitigation policies. They do nothing to reduce their emissions. Here all three components of attitudes are also largely consistent.

Figure 5.1 shows the distribution of the six segments in October 2014.

While the Alarmed and the Dismissive have attitudes that are largely consistent in their components (thinking, feeling, doing), the groups in between do not. For three out of four people, feelings, knowledge, and behavior about climate change don't really match up.

The hope among the climate scientists has been that providing more climate information could move more people into the alarmed group. This hope has been reinforced by looking at the strength and weight of the scientific evidence: It's so clear! Obvious! And the scientific consensus is so

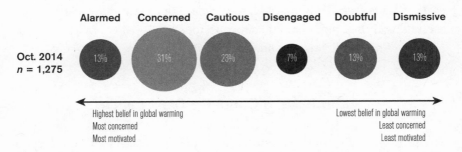

**Figure 5.1.** The "Six Americas" of climate change attitudes in the United States. Source: Yale/George Mason University.[8]

strong. But attitudes toward climate issues are not changed by information alone. One reason why just sending more climate information to middle groups such as the Cautious, Disengaged, and Doubtful doesn't really change attitudes is because new information almost exclusively targets the cognitive component. That's just one out of three components that make up an attitude. Most people have by now cognitively understood that burning fossil fuels causes global warming,[9] but what is the link to the other two components, emotion and action?

Newspapers, magazines, and other media have over the years used a lot of photos of polar bears and melting ice to try to add some feeling to the numbers, and thus to address the "affect" component of attitude. For a while the images of lonely polar bears with sad faces on sinking ice served an emotional function in our cultural subconscious. However, after prolonged overexposure, we've become comfortably numbed. Other images, too, have become stereotypes of this difficult-to-visualize issue: disappearing glaciers and photos of extreme storms or massive floods and large smokestacks. All grim imagery.

The lack of attention to the emotional component of climate attitudes has come at a very high cost. The bleak message that's being repeated is subconsciously being heard as an accusation: Not only is the human influence on the climate system too high, but filthy-stinking-rich you are part of that problem, killing sad and cuddly polar bears, melting away pristine ice ecosystems, and subjecting poor people to harm and injury. This may factually be correct, but the point here is that today, the condemnatory tone of climate communication has created strong associations with fear and guilt. Climate images and messages indirectly underscore that I should feel bad about the way I live: *My ecological footprint is way too high for the earth. Fragile ecosystems are collapsing under the weight of my lifestyle.* The message—at the emotional and subconscious level—is: *Clean up your act or we'll all go to hell.* It is, in one word, shaming.

## Attitudes and Cognitive Dissonance

The problem with using fear and guilt-inducing messages and imagery to influence attitudes is that it tends to backfire. Particularly when the opportunities for a change in behavior or lifestyles are hard to realize. For instance, many respond positively to the question of whether they would like to buy greener products. But very few are actually able to act on that intention due to a whole range of factors—from lack of available options, poverty, and confusing eco-labeling to lack of social support, time, and knowledge. That's when frustration sets in. We feel lousy—squeezed between a rock and a hard place. The ways we think, feel, and act about the issue just don't go well together. Therefore, something must be done about this unpleasant message.

Being human, these unpleasant feelings are not hard to resolve. Our creative minds quickly find a solution to this sorry state: *Rather than changing my actual behavior, I can modify my thinking to match what I do.* We've learned how to do this before with other issues: In social psychology, this way of dealing with our attitudes has a special name, cognitive dissonance. It has been intensively researched over the previous decades.

Cognitive dissonance refers to a situation in which there are conflicting feelings, thoughts, and behaviors. It is an uncomfortable inner tension. It only dissolves when one or another of the components changes and harmony is restored. The research on cognitive dissonance started when psychologist Leon Festinger infiltrated a religious sect in the 1950s.[10] They were preaching that the end of the world was near. He was curious as to how they would psychologically cope with the situation if (or when) the end of the world failed to materialize on the predicted date. Would they change their beliefs? He suspected not.

The fateful day came and went. But the sect didn't give up on their beliefs. In the morning they were bewildered, and felt a range of emotions: let down, confused, relieved, or even angry and embarrassed (states of dissonance). But then the sect leader channeled a new revelation: The world had been saved at the very last moment, exactly because of the sect's steadfast belief and strong dedication. What more proof could they ask for? The world did not end, and it was all thanks to them and their actions. They had been given a new opportunity!

This was a new thought for the sect. It resolved the intense dissonance between what they had thought (the end would come), their feelings (disappointment), and their behavior (the member's vigil and shared actions for the sect). Now they could continue their preaching to the world with pride and describe how they had saved the planet—while welcoming new members into this incredible community. It was precisely those members who had been the most loyal and invested the most into the issue who were the most strongly convinced of the remarkable turnaround. Festinger observed that some of the more peripheral members drifted away.

Festinger later devised some cunning, now classic experiments[11] to see if this social-psychological adjustment of our thinking to match our actions would hold up in general. For instance, he asked a number of students to perform dull tasks, such as turning pegs in a pegboard for an hour or tediously sorting irrelevant tokens. Since the task (doing) was really, really dull, the expected attitude would have been to judge it (thinking) and consider it truly boring (feeling). But then after the participants had completed the task, he offered half of them a small amount of money to tell a waiting participant (a confederate) that the tasks were really fun and interesting. The other half, he offered far much more money to do the same thing.

Finally both groups were asked their own opinion about how boring or exciting the task actually had been. The idea was to test whether receiving a reward for telling someone else that the activity was fun had any impact on their own opinion. It turned out that those who received a lot of money confirmed that it was actually boring. They lied to the next guy for the money. That is consistent—it's a "good reason."

But those who lied to the next participant for very little money seemed to experience dissonance. They began to ask themselves why they had lied. Why had they said it was fun when it had been boring? They wouldn't admit they had agreed to lie just for a few dollars. Their creative solution to reduce the dissonance was to start thinking that the task was, um, kind of fun after all. The results showed that those who received little money turned out to have much more positive attitudes to the task. Since it wasn't the petty money, then the task itself must really have been kind of cool. *What they actually did overruled what they felt and thought.*

By itself, such lab experiments have limited validity. But the psychological dynamics it illustrates have since been repeated and extended in hundreds of different settings.

Cognitive dissonance rose to fame through smoking research, which is very relevant to climate attitudes. Smoking researchers knew that those who smoke have more positive attitudes to smoking than those that don't.

It plays out like this: *I smoke. I also know smoking leads to cancer.* These two thoughts (one from behavior, one from information) don't go well together. Particularly if I also hold the value of wanting to stay healthy and live a long life. So action and knowledge are at odds and generate the type of discomfort called dissonance. I could stop smoking. But it's hard to break the habit *and* to keep my smoker friends. If I continue smoking, then, the dissonance needs to be dealt with somehow.

It turned out that smokers in general use four main strategies to deal with dissonance. The first is to modify perception of reality: I could convince myself, for instance, that I don't really smoke that much. Compared with others, I smoke much less. The second strategy is to weaken the importance of the concern: *The evidence is rather weak that smoking causes cancer. My aunt smokes forty a day and is fit as a fiddle. But my other aunt, who never smoked, she died of cancer!* Third strategy: Add extra cognitions, such as *I exercise so much that it doesn't matter that I smoke.* This is a kind of moral license people give themselves, as in: *I ran two miles, I deserved a few scoops of ice cream.* Or *Since I've given to charity, I can now indulge in shopping with a good conscience.*

Finally, in the fourth and most radical strategy, I can deny that there is any relation between smoking and health. I can convince myself that the so-called evidence that smoking leads to cancer is just propaganda. It is an excuse for people who like to control others, who enjoy the power of deciding how others should live their lives.

Cognitive dissonance in attitudes to climate change follows the same pattern as it did in the case of smoking.[12] We have two thoughts, or cognitions: *I have a large carbon footprint.* And I've learned that $CO_2$ *leads to global warming.* These two notions don't go well together. They conflict with a positive self-image, and create a vexing discomfort.

In order to deal with the dissonance, the exact same four main strategies seem to be used: First, I can say that my footprint is really quite insignificant. Depending on where I live, I can say it is the emissions from some faraway spot—be it China, the United States, the European Union, or Russia—or from oil companies that are the real problem. *It's not me, it's them! I'm not implicated.*[13]

Second, I can reduce the importance of one cognition by starting to doubt it: *Well, the evidence is quite weak that $CO_2$ really causes warming. There hasn't been any warming since 1998. It seems to be the sun really.* Or: *The alarmists are exaggerating. I'm cool and reasonable.* In this way I can tell others, if asked, that I may be somewhat concerned, but the issue is unimportant. This is the mind-set of the doubtful segment.

Third, I can add cognitions to make me feel better: *I've installed a heat pump and switched to LED lightbulbs at home, so now I can fly to Thailand with my family with a clean conscience.* This is the sweet side of green consumption, as moral license: *I can buy a few green products as a way to reduce my dissonance—and then indulge in other major purchases.*

Finally, and most drastically, I can dismiss and deny the whole thing: *There is no real evidence linking $CO_2$ and climate change.* Or, *Oh, it's all just BS!* Or, *The IPCC is nothing but another doomsday sect.* Or *It's all a conspiracy and a hoax.* Or, *Jesus will be back before the world ends.* This is the mind-set of denialism in those who dismiss the whole notion of climate change.

In most of us, the global warming message initially evokes troublesome feelings, such as uneasiness, fear, or guilt. The more we believe the message, the worse we feel, as long as we don't change our behaviors. Failing to act on what we know just increases the dissonance. We struggle with this internal conflict, and start negotiating with ourselves.

From this unpleasant dissonance arises the wish—a market demand, you could say—for ways of reducing it. This is the demand-side pull that makes the doubt, indulge, and deny messages an easy sell. Weakening the climate message makes us feel better about ourselves and the high-carbon lifestyles we all have.

This doubt then becomes a major barrier to widespread acceptance of the climate message. Climate hell doesn't sell. Or maybe it did—in the beginning, when it was news. But now there has been a gradual backlash against the climate message, since we're stuck with high-emission lifestyles from the previous century. Dissonance has, year by year, been grinding away on the cognitive component, weakening it to make it more consistent with our actual behavior and helping us restore our feelings about ourselves as good people.

For my own part, I feel dissonance each time I fly. I still do it, though. I'm far from a climate saint. It doesn't help much that I use my electric bike as much as I can when home. My own solution is to buy four times the amount of carbon quotas that I fly for, from the EU trading system. If I want to participate in our current society, and contribute to the swerve, I'll have to endure some inner dissonance. There is no such thing as 100 percent consistency or 100 percent purity. We're all in this mess. (See!? I'm hard at work here to reduce my own dissonance by explaining it away.)

What this well-researched theory tells us is this: As long as there are few opportunities for simple and easy climate-friendly behavior, and the message stirs feelings of fear and guilt, then this dissonance will generate a slow, grinding backlash against the climate attitudes.

Consider what happened in 2007, when there was huge media coverage of the IPCC report, the film *An Inconvenient Truth* was released, and the Nobel Prize went to Al Gore and the IPCC. Those alarmed and concerned rose again to 63 percent by 2008. But then the ensuing Copenhagen negotiations failed to create public policies, the financial crisis hit, and the opportunities for consistent behavior never materialized. All this increased the cognitive and social dissonance among many who had been concerned about the climate. And Climategate came along to manufacture doubt about the honesty of climate scientists. The dissonance could then be done away with by modifying the cognitions: *It's not that serious. Maybe the science isn't settled yet. Maybe the models are wrong. I am now less concerned about climate.* The more waiting and seeing they could do, the better people felt.

Modifying how we think about a problem is our favorite way to justify our actions. The worry and dissonance vanish like ghosts in sunlight. The inner problem is solved. Figure 5.2 shows that as of April 2014, just

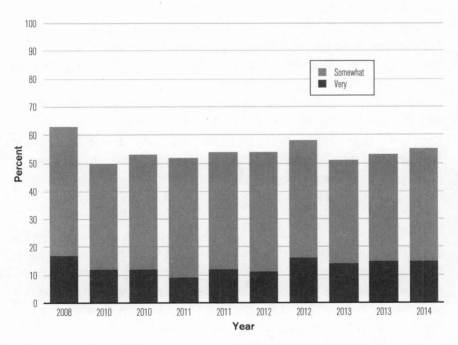

**Figure 5.2.** Americans' response in the period 2008–14 to the question: "How worried are you about global warming?" Source: Yale Climate Change and Communication project, with around a thousand respondents per year.

15 percent of Americans were very concerned about climate change and 40 percent were somewhat—a decline from 2008. It appears that despite steadily stronger climate facts from ever more reports between 2008 and 2014, public attitudes have changed very little.

## Do Attitudes Determine Behavior— or the Other Way Around?

The crucial question in attitudes research is: To what extent do attitudes guide behavior? Many reason that a shift in attitude should result in actual behavior change. If so, then we would expect a shift in real-world behavior to follow *after* a change in thinking and concern. For instance, we would expect someone who shifted from being doubtful about climate change to being concerned or alarmed to consume less, make greener purchases, and vote with the climate in mind. But do shifts in thinking really translate into consistent behavior?

The answer is, as so often in psychology, which studies the boundless peculiarities of people, *it depends*. For instance, let's follow a vegetarian into a restaurant. The vegetarian's attitude to meat is a negative one, as we all know. Will this attitude translate through into real action when ordering food from the waiter? I know this woman well, and we sit down with our friends. The menus have such a range of options. What will my vegetarian friend pick? When taking the orders, we all wait in suspense for this exciting cliffhanger moment: Lo and behold, the vegetarian orders the vegetarian option! I predicted human behavior—despite free will and the notoriously precarious behavior of humans, including my dear vegetarian friend.

What about following a man of conservative attitudes into the voting booth? We know—based on our talks with him—that he says no to abortion and yes to guns, drives an SUV, and wants a small government with low taxes. Which party will he vote for? The greens? If you bet on that, the odds against you will be very high. Being a fly on the wall, I can observe that he actually put the conservative vote into the ballot. No surprise.

So, yes, in those cases where attitudes are deeply ingrained into our values and identity, then attitudes actually do determine—and predict—human behavior. Together with values, they trump the flimsy influences of situation and create a predicable behavior across the board. The attitude is consistent. No cognitive dissonance here.

But in other cases, situations get their revenge. Let's say that, last week, our smoking friend finally decided to give his habit up for good. He knows that smoking kills. But just now when sharing a glass of excellent wine with his best friends, in the name of community—he feels like joining them in having a cigarette. Only one. It tastes good. And feels relaxing. But after this wonderful pleasure he remembers the decision from last week. The dissonance grows inside.

Again, action and thinking don't match up. What a bummer. Our smoker might think, *What do I do? Ah, now that I've lost, I might as well have a few more.* (That's consistent.) *Don't I make decisions for my own life?* he adds, already feeling better. But finally he firms up that he'll surely quit again tomorrow or at least next week. Or he decides that this whole anti-smoking thing is propaganda and manipulation of free people by those who like to control others. In either case, dissonance is resolved, for now.

The clear conclusion, from the social psychology perspective, is: More often than not, it is actually behavior that determines attitudes, not the other way around. If our lifestyles are far from climate-friendly, then our attitudes tend to follow.

## Social Dissonance: Do I Have to Change My Friends?

Attitudes do not only link what we ourselves think, feel, and do; they also link us to other people. My friends and colleagues tend to share opinions with me. Golfers and Harley-Davidson fans each have their set of attitudes toward the things they cherish. Common attitudes are thus human bonds. And breaking those bonds may come at a high social cost.

In the article "How Politics Makes Us Stupid," Ezra Klein pointed out the deep roots of partisanship with this scenario for conservative Fox news anchor Sean Hannity:

Imagine what would happen to, say, Sean Hannity if he decided tomorrow that climate change was the central threat facing the planet. Initially, his viewers would think he was joking. But soon, they'd begin calling in furiously. Some would organize boycotts of his program. Dozens, perhaps hundreds, of professional climate skeptics would begin angrily refuting Hannity's new crusade. Many of Hannity's friends in the conservative media world would back away from him, and some would seek advantage by denouncing him. Some of the politicians he respects would be furious at his betrayal of the cause. He would lose friendships, viewers, and money. He could ultimately lose his job. And along the way he would cause himself immense personal pain as he systematically alienated his closest political and professional allies. The world would have to update its understanding of who Sean Hannity is and what he believes, and so too would Sean Hannity. And changing your identity is a psychologically brutal process.[14]

But we're not really stupid. In a way, we're rationally recognizing the costs of breaking the social attitude bonds with those that support us. Referring to a study by Yale professor Dan Kahan, Klein writes, "Kahan doesn't find it strange that we react to threatening information by mobilizing our intellectual artillery to destroy it. He thinks it's strange that we would expect rational people to do anything else." Kahan, as Klein notes, puts it this way:

Nothing any ordinary member of the public personally believes about the existence, causes, or likely consequences of global warming will affect the risk that climate change poses to her, or to anyone or anything she cares about. However, if she forms the wrong position on climate change relative to the one that people with whom she has a close affinity—and on whose high regard and support she depends . . . in myriad ways in her daily life—she could suffer extremely unpleasant consequences, from shunning to the loss of employment.[15]

So friends may support the same political party, belong to the same religion, and be of the same persuasion. However, if two people like each other but have strongly opposing views on an important issue or a third person, it may become difficult to remain friends. Either they will decide not to talk about what divides them, or they will withdraw from each other. On the other hand, if both strongly dislike a third party, they may become comrades in arms, despite their differences. It is as difficult to praise George W. Bush among liberals as it is to praise Al Gore for his climate activism among conservatives. Lashing out at Bush or Gore or Obama can make a good together-feeling when we agree. Also, if liberals attack Fox News and Hannity, then conservatives conclude Fox and Hannity must be good people. And vice versa. "The enemy of my enemy is my friend," as the saying goes.[16]

That's why people not only try to make their own attitudes internally consistent, but also try to align their attitudes with their friends' attitudes, or at least ensure they contrast with the attitudes of their out-group, their enemies. Eventually, conflicted people will try to balance their relationships with significant others by harmonizing their opinions. And that is how someone can know and understand all the facts about climate, but decide instead to trust the opinion of someone they admire and want to keep as a friend.

In summary, our climate attitudes are doubly embedded, both in an internal matrix (affect, behavior, and cognition) and in an external network (social relations). When provocative information comes our way, we employ self-justification to keep our attitudes aligned within ourselves and with significant others. And we didn't just start acting this way yesterday. Two centuries ago, Ben Franklin wrote, "So convenient a thing is it to be a rational creature, since it enables us to find or make a reason for everything one has a mind to do."[17]

—SIX—

# The Roots of Denial:
# The Psychology of Identity

For many years, I worked as a scenario-planning consultant for large corporations and government. One week, my colleagues and I were running a large workshop for a multinational oil and gas company, teaching methods for thinking more long-term. The location, near the gulf of Mexico, had a magnificent view of a freshwater lake from the workshop windows. Only the lake had gaping, gray muddy shores that dropped down more than forty feet before hitting water. These parts had experienced a multiyear drought, and the lake was showing obvious signs of severe drying. The boat ramps had become unusable. To enter the boat for a trip on the lake, we had to descend a ladder. It was still nice to be on the middle of the lake. We all got champagne. Brilliant weather. Jumping rays of light illuminating the ripples in the breeze. Sunglasses, white shirts, and smiles.

But all around us the exposed shores were quietly telling a stark story. The weird thing was, though, that there was little willingness among the more than one hundred executives to discuss climate disruption—whether looking at the dying lake (and shrinking groundwater supply) in front of them or out into the long-term future. They'd come here from all over the world to advance the company and their own career in it. But mulling over looming climate disaster due to oil production is not the best way to exhibit a go-getter attitude and get a promotion in a petroleum company.

It was nearly impossible to keep the urgency of climate alive in the strategy discussions. When we tried to raise the subject of climate risks and opportunities in strategy discussions, we were labeled "European socialists" by one of the group's top brass, not exactly a flattering label from the mouth of a Texan oil executive. As a facilitator, I felt I'd failed; we hadn't designed the process so that the participants could break through their corporate and personal resistances.

In small talk over dinner, and while in taxis together, what you might call the professional petroleum mind-set came out clearly. When I asked any one of them about the drought and climate issue, answers went something like this:

What the last IPCC report actually showed was that global warming isn't happening as expected. There is much less warming, a pause. So it is clear that the scientists were wrong and now have to go back to their models. They predicted the hurricane season was going to be bad, but then we had hardly any hurricanes this year. It's a long time since Hurricane Katrina now. You know, these climate guys are just against everything. But Al Gore has this big mansion, twenty thousand square feet or so, and flies his own jet. He's a doomsayer, yet he invests and earns a lot of money. And those environmentalists are against all kinds of energy. Before, the environmentalists were in favor of more natural gas to replace coal. But now, they are against gas as well? And they are even against windmills, they say, because they kill birds! Come on! [Laughter.]

In this reasoning, we can see how the mind works. We all work and live based on the fossil energy that fuels our society. Along come some climate scientists who declare that rising emissions from our current economy will in the long term destroy the ecology our civilizations depend on. And even worse, they indicate that government-enforced reduction of fossil fuels is the solution. It doesn't matter whether they call for regulation, taxes, or consumer change; all such policies are a direct challenge to the petroleum and libertarian worldview.

Those faced with such a challenge feel a need to defend their identity and lifestyle against the message that climate disruption is real, urgent, and caused by human fossil fuel use; they feel an inner need to explain it away. This defense can be achieved by targeting the messengers: "They're obviously wrong." "They are cunning and deceitful." "They're nuts." "They are gloomy killjoys." Then, to keep the good-me and good-us feeling, it is also important to see the shrinking lake as a victim of natural drought. It is not something in which I or we are complicit. First, "They're wrong," and then "It's not me or us."

It would be easy to understand if this view were held just by oil workers and petroleum executives who wanted to defend their own jobs. But the need to remain innocent is much broader, extending to the many who feel that their way of life and core beliefs—perhaps in a free market and or in personal choice—are threatened by ambitious climate policies.

In previous chapters we've identified several defenses to climate messages. First, we saw how humans have evolutionary traits that tend to make climate seem distant and low on the urgency scale. Then we saw how catastrophe, loss, and cost framings backfire—and how our habitual actions create dissonance that lends itself to doubt and denial. In this chapter, we'll explore how cultural identity helps determine what passes as fact and risk, and how changes in self and identity might happen.

It's perplexing when and why climate became a politically divisive subject, particularly among white males.[1] Back in 1990, the topic was not so clearly drawn into the culture wars of the United States. Democrats and Republicans did not have a big divide over climate change, only a gap of around 10 percentage points in 2001. Gradually the topic became more political, and views about it started to become partisan, and connected to cultural identity. Then, by around 2010, polarization had inflated the gap from 10 percent to 41 percent; with 29 percent of Republicans and 70 percent of Democrats who believed climate change is already happening.[2] The widening of this gap has in a large part been due to Republican party activists, in collaboration with fossil fuel interests funding conservative think tanks. This has been accompanied by a well-funded campaign claiming that jobs will be lost and the economy will suffer if climate change measures are introduced.[3]

Hence, in the minds of many, climate science is now strongly associated with "liberal views" and "the left-wing media."[4] Also, because Al Gore and many public liberals have been strongly supportive, the inherent dynamics of partisan politics ensures that conservatives must be against it. Now, if you know someone's view on gun control, abortion, and same-sex marriage, you can pretty well predict their views on whether global warming is real and human-made. The issue has become polarized.

## The Cultural Cognition of Facts

One of the best-documented findings in psychology is the so-called confirmation bias. This means that we automatically look for information that confirms what we already think, want, or feel. If asked "Is Roger friendly?"

we start to scan for friendly memories of Roger. If I hold a conviction that left-handed people are more creative, I'll take special notice of all the creative left-handed people that I meet, and ignore many of the creative right-handed ones. We tend to search—in our minds and around us—for confirming evidence. Opposing information is filtered away.

The problem is that this sometimes leads us to avoid, shrug off, or forget vital information that would require us to change our own beliefs and behavior.[5] In terms of climate, the bias works both ways: Those who already are concerned about global warming will read more news that confirms it, and those who believe it is bogus will prefer news sources that question the science.[6] The bias is reinforced when the beliefs are connected to deeper values and identity. People then exhibit a strong selection preference for information on climate change that already matches and reinforces their cultural identity and worldview. Once the polarization gets going either way, the confirmation bias makes it self-reinforcing.

We prefer experts and information that match the values in our own cultural group, and easily dismiss those of opposing groups. This way of thinking has been termed motivated reasoning and cultural cognition in the research on climate risk perception. We don't take scientific facts at face value. We check who the expert messengers are; do they hold values and a worldview similar to ours? If yes, then the expert is "good" and we're predisposed to agree. If not, we'll question their methods, analysis, and results. Kahan and colleagues from Yale have performed several studies that confirm the role of cultural cognition on our perceptions of whether climate change is real and urgent.[7]

People with conservative values (often called hierarchical and individ-ualistic values) tend to be skeptical of environmental risks and to dislike regulations to limit those risks, since regulations restrict free choice, commerce, and industry. On the other hand, liberals (who hold egalitarian and communitarian values) more often view commerce and industry as self-interested and polluting entities that create unjust disparity. Since they tend not to like commercial risks, they feel stricter regulations are justified.

Kahan's team found that a solid majority of liberals, 68 percent, held the view that most expert scientists agree that global warming is real and human-caused. Only 20 percent believed scientists are still divided. On the other hand, a majority of conservatives, 55 percent, believed that most expert scientists are actually divided on whether humans are causing global warming. And yet another 32 percent perceived most expert scientists to fully *disagree* with global warming. This adds up to 87 percent of conservatives but only 32 percent of liberals saying that climate scientists still are divided about global warming. There is a gap of more than 55 percentage points in the two groups' views of the exact same scientists.[8]

The point is, whether people believe in climate messages from scientific experts, like whether they believe in evolution, says more about their cultural identity than about the science.[9] Put differently: Many conservatives don't oppose climate science because they are ignorant. Rather, it is a way of expressing who they are.[10] This obstacle becomes the innermost barrier to climate communications: The messages crash against the wall of the self.

In one of the Yale studies, participants were asked how knowledgeable and trustworthy they thought different experts were. The experts they were asked to rate were fictional. They had the same academic credentials but contrary cultural values. When the participant agreed with an expert's cultural values, the expert was seen as much more knowledgeable and trustworthy than when the expert held the opposite values.

This tends to lock messengers into pre-sorted boxes. Let's say I see that Al Gore, Bill McKibben, James Hansen, or Naomi Klein is going to say something. Even before they open their mouth, it's already decided what I'm going to think about their message, whether I'm pro or con. The same applies to messengers such as James Inhofe, John Boehner, and Sean Hannity. My filters predispose me to whatever words they utter. This greatly diminishes their potential to reach out beyond those with similar cultural values. If an oil analyst, doctor, or entrepreneur said the same words, they would resonate differently.

Other psychologists and sociologists have similar findings.[11] The strongest predictor of expressing climate change denial is having a libertarian, free-market worldview. Or as psychology professor Lewandowsky put it in an interview: "The overwhelming factor that determined whether or not people rejected climate science is their worldview or their ideology."[12] Worldviews are closely knit with identity and job or profession. If your job is in the fossil fuels industry, or you depend on cheap gas (thanks to a long car commute to work, say), it is psychologically much harder to admit that your lifestyle is disrupting the air than if you work in the re-insurance industry (which daily calculates climate risks) or drive an electric car.

If we're dealing with problems where the solutions (higher carbon taxes, for instance) could threaten our tribe—or, worse yet, our own social standing in our tribe—we tend to rationalize. The obvious consequence is that, when communicating climate science, we must find more communicators whose cultural identities are consistent with those of their audience. A good example is Katharine Hayhoe, who is both an evangelical Texan and a climate scientist. Living in Texas, she can engage locals in dialogue in a way no outsider can.[13]

## The Search for a Good Reason to Reject Good Science

The climate consensus came from thousands of scientists independently arriving at the same view based on many, many unrelated sources of evidence. If you want to reject an overwhelming scientific consensus, and keep a positive self-image, you need a good reason for knowing better. And that need launches a search for supportive evidence, no matter how small, to bolster the opposing view.

People in this situation can be willing to believe any self-proclaimed expert who espouses the same cultural values. Or they can demand more and more evidence from the other side, never being satisfied, since they do not ask if their own filters and core beliefs might be wrong.[14] When supporting facts are few and far between, it then becomes easy, perhaps even helpful, to develop a symbolic reality—to welcome a conspiracy theory about why the facts must be wrong.[15] These archetypal fantasies might depict the denier as a Galileo-like hero "who opposes a corrupt iron-fisted establishment"[16] that is trying to bully truth-speakers into silence. Believing in such stories serves a purpose: It creates group cohesion.[17] And then it becomes hard to stop believing in them.

In fact, once people have identified with their position, providing contrary evidence can actually make them more resolute in their convictions.[18] Remember those who interpreted Climategate as confirmation of their belief that scientists were hatching a huge, nefarious plot? In their eyes, the three or four reports that subsequently vindicated the scientists and the science proved only that the circle of conspirators was wider—and more dangerous—than previously suspected.[19]

Even experienced scientists, when engaged in a particularly nasty internal ideological conflict, have been known to deny the science at hand.

It takes hard work to force yourself to override your cultural identity and face facts. Some are able to. But for the most part, when identity and fact conflict, we end up with what you might call smart idiots—people who use all their intellectual capacity to defend a false view so their identity is unharmed. But it is a sort of pseudo-stupidity, since there is lots of intelligence at play. Well-designed studies show that the better science literacy someone holding a conservative ideology has, the more wrong she will get the climate science.[20]

This is a form of human self-destructiveness that is well known to psychotherapy: Sometimes the inner resistance will come up with no end of explanations and excuses as to why we keep doing what is not good for us. Very bright patients can become especially challenging to therapists, since they make therapy into a competition, analyzing the therapist's comments to prove him wrong or incompetent. Some use their intellectual capacity to bolster their defenses until they are unassailable.

## Hiding from Ourselves: Lessons from Psychotherapy

In public climate debate, the word *denial* has become a word of accusation, hurled toward political foes. Calling someone a denier has, particularly for the left half of the political spectrum, become a way of defining the other side, their out-group. Thus, *denial* may by now have become a hopelessly muddled word, too overused and unclear to be useful, having been debased to a label of "ignorant" or "stupid." But rather than condemning denial, we can choose an attitude of curiosity. To do so we can draw on some insights from psychotherapy.

Psychotherapy is really an intensive study of the processes of personal growth and identity transformation. After a century of practice, it is clear that deep personal change doesn't come easy. Many choose to enter coaching or therapy because they really, really want to change. Yet they soon enough run into an inner resistance. It's as if there is something inside, a shadow or a darker twin, that refuses to be happy about change. This stubborn resistance may make life miserable, barring progress on procrastination, smoking, drinking, other addictions, trouble with romantic relationships, or whatever else is at hand. Something self-destructive inside obstructs all good intentions to improve—particularly when told to by others. All schools of therapy from early psychoanalysis through to modern cognitive behavioral therapy have had to deal with this phenomenon.

Resistance shows itself as automatic reflex to maintain a sense of security and avoid discomfort. We unconsciously protect convictions about ourselves and important others. Psychotherapists might hear a patient say, "Even if my husband does get violent sometimes, he really is a good man." Resistance knows no limits when it comes to tricks, ploys, dodges, deceits, or beguiling detours. It can draw on any person's best intelligence, creativity, and emotions—all in the effort, usually, to protect against inner vulnerability and anxiety that change will bring. It is hard to beat, even on a good day.

Therefore, most coaches and therapists today prefer to view patterns of resistance as a portal to understanding the client better. They emphasize the importance of working *with* the resistance and not *against* it. I'm not suggesting we put climate deniers on the couch. But I am suggesting that we might apply this common therapeutic notion to climate communications.

Maybe it would be better if we stopped speaking of climate denial altogether, and rather spoke of resistance. Resistance is a less accusatory and more inclusive word. Being labeled a denier never helps anyone change; it only fires up the resistance. I, too, can feel resistance, if I really take in the full implications of global warming. After all, the climate messages are threatening, apocalyptic, and overwhelming—everything that awakens our inner resistance. Taking them seriously means considerable changes in our outlook and lifestyle.

It may be frustrating for climate scientists that climate change messages engender such resistance, but for psychologists it's not the least surprising. There are no simple solutions to global warming. And the forecasts are dark. An easy way out of the anxiety is to not want to know too much about what is coming. Thus, to some extent, we all resist taking in the full ramifications of the climate disruptions.

If we let the resistance develop into full-blown denial, this leads to a splitting, an inner wall.[21] Unpleasant facts are stowed away in some concealed part of the mind, far from what we access in our everyday lives. Certain thought patterns must then be forcefully held on to in order to maintain the splitting: "Those warming alarmists have *obviously* gotten their models wrong." "It's a leftist fraud." "This drought is natural—everything goes in cycles, and there was a pretty bad drought in the 1950s, too, and in centuries before that." "This wildfire is part of nature—there have always been fires." And so on.

This dynamic goes some way toward explaining the venom and malice that is observed in the climate debate, particularly in Internet comments. Climate deniers are eager to hit back at those who do not live up to the principles they're perceived as preaching to others. This may be climate scientists who fly to far-off climate conferences ("see—they're dishonest, deceitful, fake, double-dealing, fraudulent"). Or EU climate commissioner Connie Hedegaard is derided for bad climate accounting in EU institutions. Al Gore is discredited for his big mansion. These are self-protecting maneuvers from a vulnerable self that somehow feels attacked. The knee-jerk reaction is to kick back, taking all the pent-up frustration that has accumulated behind that inner wall, and throw it out at the personified projection. Attack is the best form of defense.

Psychotherapists know that if you push too hard against the inner wall of denial, all you do is strengthen it. Poking the projection mechanism too early backfires on the therapist. The self works hard to undermine whatever it perceives as a threat, and identity easily eats reality for breakfast. We humans are prone to deny a host of things to protect our self-esteem: infidelity, shoplifting, lying, envy, slapping, cheating, whatever. Deny, and then cleverly deny that we are denying. Sometimes we get so good at it, we hardly notice what we're up to.

In the case of climate, the individual denial is strengthened by the social and corporate organization of denial. Indeed, this is the purpose of funded denialism. It is an intentional effort by anti-climate policy groups to throw fuel on the fire of identity-protecting denial among citizens.[22]

## When Will We See a Swerve in Cultural Denial?

Denial is, unfortunately, nothing new. All cultures depend on individual and cultural denials for a certain set of topics. For a long time in Western countries, slavery was accepted as normal, and the suffering and human rights of the slaves were culturally denied. Likewise, women had no right to vote in many democracies until well into the twentieth century. Politics was for men only. And indigenous peoples were moved around like cattle by white governments. There was a shared and unspoken agreement among the ruling classes not to touch such topics. If someone spoke out against it, they were ignored, or told it was economically necessary, historically normal, and therefore natural.

Pop psychology understands denial as something we should acknowledge, rise above, and be done with. We hear people say, "You're in denial about your drinking." Or, "You're in denial about your underlying anger. You need to own it and grow out of it." But this popular view ignores the fact that every personal life and every society is built on denial to a certain extent. We cannot become conscious about every aspect of our lives. Denial and self-deception are part of being human. As the poet T. S. Eliot wrote: "Humankind cannot bear too much reality".[23]

Whether we are talking about slavery, voting rights, malnourished children, torture, racism, or industrial meat farming, the border between what type of suffering is seen as normal and what is outrageous is a line that moves over time. It shifts as individuals interact with culture and "negotiate" with fellow citizens. Thus, it is a matter of social justice to determine which forms of denial really matter. The important question is not really Why do we shut out the dreadful injustice and suffering? but Why do we ever *not* shut things out? The key to breaking cultural denial is to discover the conditions under which information is first acknowledged and then acted upon, says sociology professor Stanley Cohen.[24]

Cohen's research shows that denial can sometimes endure a long period of normalization only to break up in a sudden moment of acknowledgment. A white South African woman broke with apartheid denial when white militia almost killed her house assistant. She explained her turnaround like this: "Enough is enough . . . I couldn't keep quiet any longer . . . This was more than I could take . . . I couldn't live with this any longer." At some point people become committed, driven, unable to return to their old lives or shut their eyes again.[25] Breaking cultural denial involves moving from information to knowledge and then from knowledge to acknowledgment. Such a break always involves emotions that bring about a shift in the organization of personality, and thus a subtle shifting of identity. Acknowledgment is the death of denial, the awakening to the real. And since our attitudes are embedded in a network of social relations, speaking out openly is not just an individual psychological act. It is also a political engagement.

Californian R. L. Miller says that for a while he didn't know much about climate change. But that abruptly changed in 2008: "I began hearing about global warming. I started researching the issue and a little while later, a wildfire hit near my house in southern California. It made it personal. This was affecting me. It was affecting my children. This was affecting my community. So that's why I felt so compelled to speak out about climate change."[26] Then he started the political action committee Climate Hawks Vote, which supports Senate candidates who make climate change a top-priority issue—those who understand that climate change is not just another political issue, but rather the greatest challenge facing this and the next generations of humanity.

Despite cases like the South African woman and Miller, there is no psychological reason to assume that simply having more catastrophic floods and storms and droughts will cause more shifts in cultural identity and denial. Witness the short-lived effects on policy of droughts or even freak storms such as Katrina (2005), Sandy (2012), and Haiyan (2013).[27] These external events are not sufficient to bring about a cultural shift. Along the way, a few brave souls may break free of denial on their own. But it takes courage for any of us to backtrack and change a position that we've lived with for a long time. And it can be surprising to learn what triggers such individual changes. As Cohen points out, we are moved not by anticipated factors, but by the utterly unreasonable, unpredictable, and even bizarre.[28]

What is needed is the work of a cultural movement similar to the ones that dismantled apartheid, abolished slavery, or took on nuclear arms. What remains unpredictable is *when* this steady, seemingly ineffectual, exhausting work will result—or co-evolve—into a seismic shift in cultural denial. But one day after the swerve, what used to be resisted and denied becomes a new, shared reality.

This is the hope—that like many of the previous cultural shifts, a dogged and persistent attempt to soften and dissolve the resistance against climate science will eventually prevail. This is a systemic, nonlinear shift, not one driven by clear causality or individual willpower. One day we'll wake up and find—to our surprise—that "everyone" else is marching, too. Enough is enough.

# SEVEN

# The Five Psychological Barriers to Climate Action

By now we've seen how easy it is to distance and doubt ourselves away from climate facts. Particularly since they have been presented in abstract, doom-laden, fear-mongering, guilt-inducing, and polarizing ways. We seem to have a rich repertoire of ways to avoid changing the behaviors that belong to our sense of self. We're clever at guarding ourselves against messages that we don't really want to hear.

Scan any given day's media and Internet coverage of climate, and you'll see all those modes of distancing and self-defense on display. For more than three decades a host of messages from well-meaning scientists, advocates, and others have tried to not only bring the facts about climate change home but also break through the wall that separates what we know from what we do and how we live.

But the messages are not working, sometimes not even for the most receptive audiences. This qualifies as the greatest science communication failure in history: The more facts, the less concern. Over the last twenty years, the messengers have encountered not only vicious counterattacks but also what seem to be impenetrable walls of psychological backlash or indifference. And in response to a sense of futility the messengers are, understandably, growing despondent and exasperated.

First we need to see through the mind-and-message clutter that has gotten us where we are today. If that seems a big task, consider this: Everything discussed in the prior chapters, everything you see on the Internet or in your own friends' and neighbors' reactions to climate change, can be condensed into five main defense barriers that keep climate messages away. They work as invisible defense walls inside that block the messages from leading to meaningful response and action.

Let's call them, for easy reference, the five D's.

1. **Distance.** The climate issue remains remote for the majority of us, in a number of ways. We can't see climate change. Melting glaciers are usually far away, as are the spots on earth now experiencing sea level rise, more severe floods, droughts, fires, and other climate disruptions. It may hit foreign others, not me or my kin. And the heaviest impacts are far off in time—in the coming century or farther. Despite some people stating that global warming is here now, it still feels distant from everyday concerns.

2. **Doom.** When climate change is framed as an encroaching disaster that can only be addressed by loss, cost, and sacrifice, it creates a wish to avoid the topic. We're predictably averse to losses. With a lack of practical solutions, helplessness grows and the fear message backfires. We've heard that "the end is nigh" so many times, it no longer really registers.

3. **Dissonance.** If what we know (for instance, our fossil energy use contributes to global warming), conflicts with what we do (drive, fly, eat beef, or heat with fossil fuels), then dissonance sets in. The same happens if my attitudes conflict with those of people important to me. In both cases, the lack of convenient behaviors and social support weaken climate attitudes over time. But by doubting or downplaying what we know (the facts), we can feel better about how we live. Thus, actual behavior and social relations determine the attitude in the long run.

4. **Denial.** When we negate, ignore, or otherwise avoid acknowledging the unsettling facts about climate change, we find refuge from fear and guilt. By joining outspoken denialism and mockery, we can get back at those whom we feel criticize our lifestyles, think they know better, and try to tell us how to live. Denial is based in self-defense, not ignorance, intelligence, or lack of information.

5. **iDentity.** We filter news through our professional and cultural identity. We look for information that confirms our existing values and notions, and filter away what challenges them. If people who hold conservative values, for instance, hear from a liberal that the climate is changing, they are less likely to believe the message. Cultural identity overrides the facts. If new information requires us to change our selves, then the information is likely to lose. We experience resistance to calls for change in self-identity.

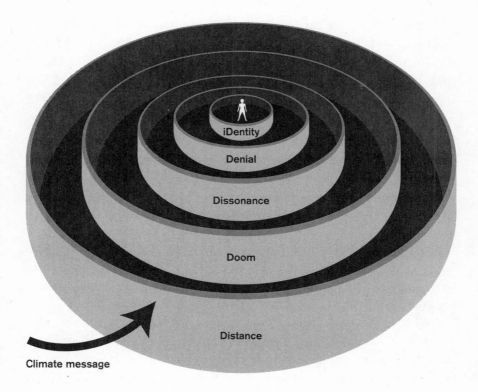

iDentity

Denial

Dissonance

Doom

Distance

Climate message

**Figure 7.1.** The Five D's: There are five barriers that block the climate message—preventing it from attracting enough concern to make climate a high priority. Crafting climate messages that work requires navigating around these five defenses.

These five barriers, or Five D's if you will, are all substantial and unyielding. Taken together they may seem invincible. They are interrelated, but still distinct. Think of them as concentric circles around the citadel of the self, with *distance* as the first line of defense and *identity* as the final, innermost defense—as depicted in figure 7.1.

The anti-climate movement has been successful in triggering each of these barriers in its battle against climate science. But inadvertently, climate communicators have activated them, too, for instance by conveying climate facts through abstract graphs and long time lines, using framing that backfires, not linking risks to opportunities for action, relying on bad storytelling, and provoking self-protective and cultural cognition by unnecessary polarization.

Knowing what the barriers are, though, and deciding what to do about them are two very different things. We've already tried breaking through them with ever more facts and eight-hundred-plus-page reports. We've gone down that road, and repeatedly found ourselves in a hole.[1] The combined effect of the five D's guarantees failure. If we find ourselves in a hole, it may be time to stop digging. It may be time, too, to leave behind attempts to hammer angrily away at the defenses, and stop blaming the other side of denial altogether. Something has to change. A different story is starting to be told. A different result is waiting to happen.

There may be ways to simply get around and beyond the Five D's. Good coaches rarely attack the habitual defenses head-on, but look for opportunities to do something else. Remember the infamous Maginot Line? The French created this heavily fortified line of defense along their border with Germany prior to World War II, hoping it would keep them safe from German invasion. But the mobile German army simply evaded the defense and went around it. They invaded France through Belgium instead.

What can anyone concerned with climate, but frustrated with the progress, learn from this? That perhaps sometimes it is better to evade defenses than attack them outright. It's futile to believe that it's possible to run right through defenses with ever more pointed and grim rational-information campaigns. But communicators *can* outflank the fortifications behind which many protect themselves from climate reality. To regain momentum, it is better to find territory that lets us move more freely around the Five D's. I propose we abandon linear antagonism, us versus them. Instead, let's move more with the flow of the human psyche. People have to *want* to live in a climate-friendly society because they see it as better, not because they get scared or instructed into it.

In part 2, we'll explore how to use what we know to create smart new strategies for climate communications and action. There's nothing so practical as a good theory.

# PART II

# Doing

## IF IT DOESN'T WORK, DO SOMETHING ELSE

*Even if I knew that tomorrow the world would go to pieces,*
*I would plant my apple tree.*
—ATTRIBUTED TO MARTIN LUTHER

# EIGHT

# From Barriers to Solutions

I had been walking through the vast Sarek wilderness of northern Sweden with my brother and a friend for more than eight days, hardly seeing any other people until we came to a small, quiet, but welcoming spot by a large lake. It was a location built by Sámi, indigenous Scandinavians, and one of their round, friendly, peat-covered huts graced the scene. Kindly invited to, my brother and our friend decided to sleep inside an old peat goahti. But I took my backpack and walked farther up the shore, across some open marshes, and toward some higher ground up the valley side.

The Arctic summer night was coming—one without darkness. The silence was utter and complete. Then two large owls appeared overhead and started circling. Not a whiff of sound. Just dark silhouettes against this far northern amber sky. I froze in my steps. Here I was, days of walking away from the farthest hint of the modernity, in a landscape as pristine as anyone today can imagine nature to be. And I was being blessed by the sacred slow dance of these majestic, huge, feathered creatures moving in a ritual circle over the marsh fields, again and again. They glided off as unannounced as they arrived.

For a moment I was released from civilization; I forgot all about economic and climate concerns. The economic machine felt far, far away. But it wasn't, neither in time nor in space.[1] Still, I felt outside it all, and could look at my home life from the outside, so to speak.

I didn't create our culture's rather absurd techno-economic system. You didn't, either. Still, we cannot just opt out even if we would like to. This is possible only for short periods, when invited to by some owls, or by the wild moon hanging, some eerie night, over desolate rooftops. I'm not an owl in northern Sweden, even if part of me wanted to be one in that moment.

Our daily options are given, guided, and determined by "the machine." I never asked for huge industrial corporations, nor for several billion barrels of crude oil. Yet I've got it around me. And even if I do participate by buying soda, canned pork, and plane tickets, there is very little reason to feel guilty about either fully embracing the machine—or fully rejecting it. I saw clearly—in that moment—how it subsumes and socializes us.

For most of us, it's not an option to opt out and stay forever in the wilds. So we are forced into continuous dissonance by the pain of knowing about the machine's violence even as we live with it. As we gradually adapt to liking and even loving it—with all the beautiful perks, from our coffee machines to our smartphones—the pain recedes into the unconscious. But then, to add insult to injury, the guilt arrives: *I shouldn't drive the car or eat meat. I should be self-sufficient and live off-grid, off the land.* The list could go on and on. But this guilt is both inflating the *I* and crushing it at the same time. Somehow we heroically put it onto our own frail shoulders to reverse the workings of the huge machine, each having to be "the chosen one."

Many environmentalists, in their communication strategies, have pointed out that we are each and all responsible through our modern consumption for the problems we face. But taking individual responsibility for stopping what our economic system collectively is doing to the climate is not just insufficient as a solution—it is also way too much for anyone to bear. Such exhortations bring on the dark shadow of helplessness. Only a very few are capable of opting out, going off-grid, off-gas, off-everything, living as free as night owls, outside it all.

Since my meeting with the two owls that night, I have pondered and studied how we can find climate solutions that go with the flow, as effortlessly as those birds flew overhead. With elegance and a quiet joy. Just so. And then, how to weave together those solutions—not alone, but with others.

The field of climate psychology can point to solutions that go with our flow as human beings. Many think of psychology as individualistic, and assume that a psychology of climate solutions would be about what each of us as individuals can do separately. That we only get better one by one. But we've now had a hundred years of individualistic psychotherapy and the state of the world is clearly getting worse.[2]

Consider, for instance, the reuse of plastic shopping bags. Like recycling, this has become an iconic sustainable behavior. Surely it has benefits, but it is not an especially good way of reducing emissions, since it remains within the individual small-actions approach. When a fee was instituted for single-use carrier bags in Wales, some psychologists at Cardiff University analyzed the impact. What they found was not very encouraging. Although bag usage went down, other low-carbon behaviors among the general public didn't pick up.[3] By itself, fewer plastic bags would seem to be a good thing. But mantras such as *Every little bit helps* can be misleading when it comes to climate change. Sometimes doing a little bit—like cutting down on plastic bag usage—simply serves to relieve cognitive dissonance for larger excesses, like an extra plane trip to Thailand.

Like all simple and painless behavioral changes, the value of bag reuse hangs on whether it can act as a catalyst for other, more impactful activities—such as truly green purchases and more vocal support for greener policies. Individual solutions are insufficient or even counterproductive unless they contribute to structural changes, too.

## The New Psychology of Climate Action

So, do we start a 12-step program for recovery of individual petroholics? No, the new psychology of climate is less about what each of us can do to solve the problem and more about liberating us from some of the most debilitating side effects of the global climate messages. Messages telling me that I am to blame. That I must give up on all shopping and flying and otherwise reduce my sinful consumption before I can speak out about it. That the only answers involve sacrifice. That it is too late. And, finally, and devastatingly, that anything I can do doesn't matter.

All such individualized perceptions generate resentment and inner counter-reactions: *I don't like being shamed by holier-than-thou, know-it-all-guys. Leftist propaganda. I hate loosing my freedom to travel. Green solutions are too costly. I can't afford it. Why bother? F\*\*k it*. And so on.

The new psychology of climate is more about how to stop banging our heads against the psychological barriers identified in part 1 in order to move beyond them. We know that the climate paradox at the individual level can be explained by evolutionary, cognitive, social, depth, and identity psychology. And we've identified the five barriers to effective climate communication: distance, doom, dissonance, denial, and identity.

What can psychology now tell us about solutions and ways around these barriers? Before diving in, I suggest we build the new climate strategies on three principles:

1. Turn the barriers upside down.
2. Stick to positive strategies.
3. Act as social citizens, not individuals.

**Upending the Barriers.** We can view our task as one of overcoming the Five D's, or we can frame it as finding ways to circumvent or bypass them. Therefore, the first principle is to turn barriers upside down. We can jujitsu them to become *key success criteria* for new climate communications. To bypass barriers, successful climate communication should:

- Make the issue feel near, human, personal, and urgent.
- Use supportive framings that do not backfire by creating negative feelings.
- Reduce dissonance by providing opportunities for consistent and visible action.
- Avoid triggering the emotional need for denial through fear, guilt, self-protection.
- Reduce cultural and political polarization on the issue.

**Sticking to Positive Strategies.** Communicators could see a huge lift in their message if their strategies are basically positive. Whatever we do should be inspiring, be engaging, and stimulate community. A solution works so much better when people want it, like it, love it rather than when they implement it by duty, guilt, rule, or fear of punishment.[4] If we define ourselves in a fight against the others, and a desperate one, those others will start fighting back desperately. But such ditch digging only reinforces the resistance to change.

We are entering several decades in which society's great swerve will be all around us. We're in this for the long haul. Why not make it enjoyable, meaningful, funny, enriching? If it is all pull-yourself-together-or-watch-out, then no one is going to hang in there to the bitter end. We've been used to messages like "If you save your earnings and behave prudently, you'll get the benefits later." But taking action on climate should not be a puritanical effort of restraint and of postponing gratification until much, much later. Caring for the air ought to go along with celebrating life itself—saying yes to beauty, yes to pleasure, and yes to flow.[5]

**Acting as Social Citizens.** Our main power comes from acting as *social citizens*—demanding societal change—not acting separately as individuals. The solutions to curbing wasteful practices and overconsumption are systemic, large-scale, and societal. But to get there, we need many small-step solutions in the right direction. Primarily these solutions should be seen and framed as signals to others that we care, reinforcing the message that more people are coming on board, joining the flow, to support the ongoing structural, systemic shift.

As fond as some environmentalists are about underscoring the personal responsibility to act, there may be a danger that pushing changes in personal behavior as the main solution—such as meat-free Mondays, self-sustaining gardening, or getting an electric car—can also make us complacent *and less vocal for change at the political and social level*. When that happens, individual action replaces the political, rather than augmenting it.

Let's assume you've already gone green individually. You've put a lot of effort into insulating the house and installing solar panels. Your home is lit by energy-efficient bulbs. You bike to work each morning. You tend your garden in the evening. The produce in your A+ rated fridge is organically grown and sourced from farms less than a hundred miles away. Maybe you've even gone off-grid, become vegetarian, and converted your car to run on solar or biofuel. What more can one person or family possibly do? Still, for years you've felt an uneasy disconnect between your own low-carbon lifestyle and the rising global temperatures it seems to have no impact on.

Individual green consumerism can be helpful, but mostly as signals, reinforcing the emerging business, regulatory, and civic networks that are enacting the green shifts in society. Due to the nonlinear, complex ways that societal transformations happen, there is no way to know how far these signals really reach. But we do know they reach farther if these same acts radiate joy and enthusiasm, rather than guilt and resentment.

Or perhaps you've done a few things—like recycle and shop green—but have started to question and doubt and give up on the few green choices you have because they don't seem to really matter. Nothing seems to change.

In both these extremes—either the heroic where *all responsibility is heaped on me,* or the defeatist where *whatever I do has no impact anyway*—the actions are seen in a self-destructive way that undermines any ability to act at all. Either our burdens are increased to a level where they become an overwhelming all-in fight, or we flip to the opposite—apathy and indifference.

Thus the fervor for strong individual action may backfire on you and your social network. Disillusionment spreads.

But there is no need to *blame* yourself or others for not doing enough (even when you don't), nor to *shame* politicians for doing too little (even if they do much too little). Shifting the blame around never does much good; the shaming game is not where we want to be. As Paul Goodman put it: "No good has ever come from feeling guilty, neither intelligence, policy, nor compassion. The guilty do not pay attention to the object but only to themselves, and not even to their own interests, which might make sense, but to their anxieties."[6]

Rather than taking on guilt, we just need more and more of us, as consumers, to keep shifting gradually toward voting with our wallets for the greener purchases. And in our citizen role to repeatedly raise our voice for what we truly want and vote accordingly. Political parties will rarely pass legislation that is patently unpopular among voters. Only when there is sufficient and persistent push can politicians really start to implement the larger and sufficient solutions. And even get elected and reelected after they've done it. Or get ousted if they don't. To make democracy work again, the public denialism and political obstruction must come at a political cost; they must result in a loss of voter support. And climate solutions have to go large-scale and systemic to have a hope of shared future success.

The necessary technologies to achieve a low-emission society are already available today. But large-scale systemic changes only come about when enough minds have changed. That means climate change isn't an individual, technical, or environmental problem. It's a cultural challenge, with solutions at the organizational social level. The strategies to resolve it must therefore also be social. It will not happen within the framework of rational actors acting individually. Societies change when citizens start to act *together* with others. And it has to start with many at once, at many places at the same time. This is where climate psychology may help: building bottom-up support for the coming swerve.

## The Road from Here: Five New Strategies

We now know that simply sending more facts and forecasts wrapped in large and glossy campaign materials in the direction of some "universal public" is not effective climate communication. We also know that the costs of greenhouse gases are not high enough to get companies and consumers to cut. Nor does it seem likely that we will get globally high and efficient taxes in the short- to midterm future. Then what to do to continue building citizen pressure on policy makers and business?

Luckily, there is a cornucopia of alternatives, a wide range of initiatives being tried and tested today. Most of them are outside the conventional rational-information model, as well as outside the price and rational-actor model, since both of these are based on too simplistic an understanding of human nature.[7] It will take a lot of innovative thinking and a lot of stamina to continue experimenting. But the work has already come surprisingly far. With international climate negotiations in deadlock for decades, the world has lately seen a groundswell of initiatives and projects conducted by regions, cities, and corporations, as well as hundreds of thousands of nonprofit organizations,[8] nifty new technology start-ups, and place-based initiatives.

Creative people are not caving in because distant treaty negotiators cannot agree on a high global carbon price. Many refuse to sit down and wait for change to happen from the top down. Successes, conferences, communications initiatives, support groups, and fresh research are happening all over. A highly diverse climate movement is being born, as we witnessed when some four hundred thousand people turned out in New York City, and roughly two hundred thousand elsewhere around the globe, for People's Climate March in September 2014. Out of all of this prolific trial and error, across a diverse front, the plurality of initiatives are seeming to come together into a limited set of strategies. Even if no one single overall solution has emerged so far, and probably never will, the general direction seems pretty clear.[9]

The emerging range of solutions is converging into five new main strategies for climate communication:

1. **Social.** They use the power of social networks.
2. **Supportive.** They employ frames that support the message with positive emotions.
3. **Simple.** They make climate-friendly behaviors easy and convenient.
4. **Story-Based.** They use the power of stories to create meaning and community.
5. **Signals.** They use indicators for feedback on societal response.

In the pages ahead, I'll discuss these five strategies for effective climate communication for transformation as distinct from climate information for enlightenment. At the same time, climate information and the pedagogical presentation of the latest scientific facts and findings must of course continue. No one knows it all. New facts are coming out all the time.

The approach involves a shift from a sole reliance on the old, linear one-to-many communication model toward a broader, interactive many-to-many communication model that includes practical engagement. Together these new strategies hold the potential of bypassing the barriers against which the conventional approach too long has stalled.

# NINE

# The Power of Social Networks

Groups and collectives may seem the least likely place to start improving the public's responses to climate. As Sting sings in *All This Time*, "Men go crazy in congregations, they only get better one by one." Sigmund Freud and Carl Gustav Jung, the founders of psychoanalysis, were deeply mistrustful about the ways groups influence the individual mind. Other psychologists have shared the concern about group conformity and groupthink. Indeed, people do change in groups—much more quickly and thoroughly than they do on their own—and much havoc can result from group dynamics gone awry. But groups can and do foster positive change as well. And when it comes to climate, group behavior can be a powerful lever for change.

Conventional climate information, however, has targeted the individual mind as if it is not swayed by colleagues and friends. NGOs and public agencies publish countless what-you-can-do lists, all heaping the full weight of climate disruption onto the shoulders of lone individuals expected to heroically conquer the ten or more tasks with sheer willpower and idealism. The approach is well intended, but underlying the push lies an individual-by-individual assumption of behavior change. To move past the barriers, we need to engage the power of our social networks to strengthen the norms to care for the air. The question is how?

The last time you were in a hotel room, did you notice the little sign hanging in the bathroom—the one asking you to care for the environment by reusing your towels? The sign often includes some numbers about tons of detergent or water used, or $CO_2$ generated, by excessive washing. Did you do as it requested? Some do, but most don't. Could applying a little social influence turn that around?

Social psychologists Robert Cialdini, Noah Goldberg, and some fellow researchers wanted to find out.[1] They set up signs informing hotel guests that 75 percent of their fellow guests staying in the very same room had actually reused their towels. They then measured the effect of this statement. Reuse suddenly rose by more than a third compared with rooms displaying the standard be-nice message. Why? The new sign used wording that aroused or "hacked" the ancestral forces of group imitation described in part 1. We are, at our core, imitators. Researchers refer to this phenomenon as applying social norms.

## Price or Peers?

Many rugged individualists proudly state that what their neighbors do has little effect on their own habits, but most studies clearly show that this influence is rather high. Peer behavior is one of the strongest predictors of green behaviors and attitudes on issues like littering, energy, and water use.[2] Since the imitation effect is so strong, asking consumers to go green will fail unless they are convinced that many others will do the same. If we can inform people of what others usually do, and in particular of what their friends and networks do, the impact of our request picks up considerably. Most of us tend to act according to what we *believe* similar others are doing.[3]

From economics we know that higher prices on carbon and strong state regulations are efficient measures to curb climate emissions through market dynamics. "Price, price, price!" is the manifesto from mainstream economists. They and many others insist that placing a high enough price on carbon is the only key to making change happen. Since the price ultimately gets passed on to consumers, the thinking goes, they will rationally choose cheaper alternatives that are fossil-fuel-free. Politicians, though, have been slow to do what both economists and the climate community have been waiting for—to slam a price on carbon emissions. If that happened, it would be great. But we might as well wait for Godot as wait for global carbon taxes. You can't rule out a political treaty miracle, but regrettably this is wishful thinking today. I'm a skeptic.

Fortunately there are other avenues to motivate societal change and build support for climate policies in the long run. New research from psychology and behavioral economics has demonstrated broad impacts of social norms and peer pressure on behavior—and sometimes those norms can be even stronger than prices. Much research in this area has been inspired by the early work of Robert Cialdini, a marketing psychology professor from Arizona State University.[4]

Take littering, for example. Experiments by Cialdini and others showed that people visiting parks dropped more litter when there was already lots of litter there. In cleaner areas, they littered less. Other experiments introduced a person—a "social model"—who littered in front of people. This strengthened the tendency further; even more people copied the behavior of the model.[5] Other studies show that if there already is litter on the ground or graffiti on the wall in a district, people will not only throw more litter and draw more graffiti but may also commit more crimes. People adjust their behavior to fit the signals sent by their physical surroundings about what a neighborhood finds acceptable.[6]

Or take recycling. Another experiment compared two messages to people in a neighborhood: One was a moral exhortation to recycle and the other was a statement that their neighbors were already recycling. When the wording was crafted to activate social norms (the latter of the two), curbside recycling rates increased by 19 percent.[7] The results also showed that the influence of social norms in communication could backfire if used without awareness. Well-intended but naïve media may publish messages such as "Eighty-three percent of people are not recycling!" or "Three hundred million plastic bottles discarded every day!" But these communicators have missed something critically important: "Within the statement 'Look at all the people who are wasteful' lurks the powerful and undercutting power" of imitation, states psychologist Griskevicius.[8] Then the implicit social norms work in the wrong way: Since nobody else does it, why should I bother? Shaming messages backfire; positive messages reinforce positive social norms.

Or take water use. Economist Paul Ferraro at Georgia State University did an experiment with one hundred thousand households. Some got messages that compared their consumption to the average of neighbors. A second group got messages that asked them to conserve water for moral reasons: "Please don't waste water. Remember: every drop counts." A third group got only technical tips on how to save water. The outcome? Again, social comparison had the biggest effect. The reduction in this group was as large as if water prices had increased by on average 15 percent. And interestingly, the savings were greatest among customers that were the least sensitive to price changes—the largest and wealthiest water consumers.[9]

The same principles hold true for climate-changing behaviors, such as reducing household power consumption. Cialdini and colleagues gave four groups of households different reasons for saving power:

Group 1: Because it's better for the earth (sustainability).
Group 2: For the sake of future generations (your grandchildren).
Group 3: Because it pays (saving money).
Group 4: Because your neighbors do it (social norms).

Which group got the greatest reduction in power consumption? No, not the first, idealistic group. And care for our grandchildren did not win, either. Not even the third group, who had learned how much money they could save, prevailed. The ones who showed the most commitment and the greatest reductions were again those who could compare their own efforts with those of similar neighbors.[10]

To most people energy use is boring stuff, and utility bill numbers usually evoke a big yawn (or anger if the cost is too high, of course). But something happens the moment the communication from the utility is tailored for you, and your own power consumption is compared with that of people you care about. Then, suddenly, interest in and motivation for changing energy behaviors soars.

The company Opower has pushed the power of social norms by transforming boring and confusing power bills into engaging, proactive customer dialogues. The company's Facebook app (see figure 9.1) lets you access your own power consumption measurements, and compare your energy-saving performance with your friends', all in real time. The company claims that in its start-up phase from 2007 to 2013, its services saved enough energy to power all of the homes in an American city of eight hundred thousand people, such as Indianapolis or San Francisco, for a year.[11]

Other businesses, too, are catching on. The mobile app Strava lets you easily record where and how much you're biking by tracking your mobile phone's movements as you ride. As you get fitter, you can track and share your progress and challenge your friends to bike more—to work, to the store, or for weekend recreation. Even better, you can add your miles to your community's biking total and compete on a neighborhood-to-neighborhood or city-by-city level. The city of Münster in Germany has had huge success in engaging people. So far, the Münster project has attracted twelve hundred participants. In the month of April 2014, 7,702 bike rides were recorded with a total of 30,121 kilometers cycled. The competition has

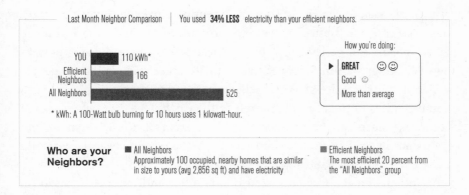

Figure 9.1. Opower's Facebook app lets utility customers compare themselves with other homes and compete with friends to save energy.

attracted 136 schools that are fiercely competing to win the most points.[12] Or you turn the app off and simply enjoy the ride.

Our social radar also influences how we view electric cars. Not long ago, their image was all square and boxy and inconvenient. Then Tesla came along, with insane acceleration, elegant design, and the most digitally advanced dashboard you've ever seen. All of a sudden, the coolest people in town are driving Teslas. They're boasting about them to friends, gossiping with one another at charging stations, and laughing at their own fuel costs compared with the rising gas prices others have to pay. The brand recognition and loyalty almost match those of Harley-Davidson bikes. In Norway, where there has been a generous tax break for electric cars, the market share has soared, making the model Tesla Model S the bestselling car of *any* type in the autumn of 2013.[13] From a life-cycle point of view, heavier Tesla models may not be as efficient as smaller hybrids, public transport, or using an e-bike, but it is beyond dispute that Tesla's launch has started to change the social norms around electric cars.

In many situations, comparison with peers is an emotional, evolutionary inner force that is stronger than rational self-interest. Facebook and Twitter users get a kick when their comments or photos receive a lot of likes. Likewise, it's no real fun just saving power, water, money, or gas all by yourself. But being seen and recognized by others for it adds zest. Cialdini points out that when people are uncertain about how to behave, they usually look around them to see what their peers are doing.

Think back to the ancestral forces identified by evolutionary psychology: self-interest, status, imitation, short-termism, and risk vividness. Social norms can be leveraged to put these forces to good use. Opower's app is asking: "Feeling competitive?—Invite more friends." Most people will not switch to low-energy behaviors because they think long-term about the fate of their grandchildren or the earth. However, when a good common cause can be associated with attaining acknowledgment and status among my in-group, suddenly the issue of climate feels near and personal, not distant and abstract.

Humans have always imitated one another and competed based on social norms. But the common cause is better served when people are imitating and competing around the common causes of water and energy savings rather than maximizing house or car size. Moreover, it appears that social norms can trump cost savings as a motivator.[14] The question is therefore not one of *either* higher energy prices *or* peer pressure. The trick lies in combining conventional policy tools like pricing, taxes, subsidies, and better technology with peer pressure to supercharge attention onto the way the policies get implemented.

So if we stop communicating our worst behaviors (how much energy we consume, how big a house we build, how many miles we fly) and instead model our best behaviors—inspired by and in competition with friends, celebrities, or anyone else we think well of—change can happen, and happen faster. We can start social cascade effects.[15]

## Social Groups: All Emissions Are Local

The term *social norms* refers to the knowledge, imagined or real, of what others would say or do in your situation. Norms are created by a broad range of social engagement, from mindlessly clicking on online petitions or imitating friends via Facebook apps to deeper engagement with identity-changing groups like 350.org or others in the Climate Action Network. The question is, Do some methods of creating social norms result in stronger impacts? Will behavior-changing prompts from Opower have lasting effects after the prompting ends? Some researchers doubt it. Rather, they say, real-world face-to-face interactions hold the potential for much deeper behavioral impact.

Since climate change is often described using highly abstract, scientific data, people yearn for personal interaction and face-to-face conversations to help them process and embody the information. Conservationists have different concerns than do health care professionals or businesspeople. Conversations with friends, neighbors, or work colleagues, or even at information booths, are far more important than is usually realized.[16] Face-to-face contact and word of mouth are still king.

What is an effective climate message? You can start telling your own personal story to people you care about. Explain why you yourself are worried about a future in which we fail to address climate change, and what kind of world you'd prefer. Resist the impulse to debate the science and be "right," and instead really listen to what your aunt/brother-in-law/ high school friend has to say. Rather than questioning the science, is this person really feeling helpless?

Another good place to start is to ask people if they are aware of the surprising agreement among scientists all over the world that global warming is real and urgent. Studies show that when people become aware of the strong consensus, this seems like a norm and they become more willing to support ambitious societal responses.[17] Therefore, simple and clear messages about the consensus, repeated often, by a variety of trusted voices, may hold potential to spread support for action through social networks.

And who is the best messenger? The most credible spokespeople for an idea or a brand are peers—people who are similar to those you want to reach and who can reach them because they are part of their in-group. Peers excel at word of mouth, a communication tool so old that cave dwellers no doubt used it to barter their animal skins for more berries. In the last few years, the advertising and marketing industries have looked to word of mouth—which today also means word of email, word of Facebook, word of Twitter, and so on—as a response to mass media overload.

As we saw in earlier chapters, people easily feel helpless if left on their own when confronted with the severity of the coming climate disruptions. Participating in a community or group that works for a common cause is a good remedy (the only one, actually) for this toxic helplessness and passivity. It can stimulate people who feel powerless on their own. But rather than forming more new climate groups and organizations that demand yet more of our free time, we can transform existing groups—sports, book clubs, churches, health organizations, housing co-ops, unions, parent associations at schools, biking clubs—into eco-teams. Make goals clear and modest, and have fun while doing it.[18]

The idea that information, innovations, and values can spread through social networks is far from new. In the field of commercial marketing, advertising campaigns targeting "opinion leaders" and influential individuals has long been commonplace. In other fields—health behavior, for example—campaigns often target peer groups and existing social networks because individuals who trust one another and pay attention to one another's behavior can effectively spread positive health behaviors within their groups.[19] Both AIDS campaigns in South Africa and anti-smoking groups in the United States have been extraordinarily successful by giving youth at risk a group to join and to identify with.[20] Why not climate communication campaigns too?

Social networks are everywhere. Friends, colleagues, neighbors, and family make up most people's network of social contacts, and they have a powerful effect on everyone's behaviors. While the required societal change has been slow at best the previous decades, today there are two new forces worth betting on: the rise and connectedness of the social web, and an increasing desire for sustainability and better self-governance among the millennial generation.[21]

Nearly all fossil emissions are made at some location, and in that sense they are local. Since they are human-made, only human groups can curb them. Acting to influence the emissions and sites nearest ourselves makes the distance less. This is a point that climate analysts tend to forget, with their tendency to focus on global averages and universal policy tools.

## Rinks, Birds, Halloween, and Cities: A Thousand Carbon Conversations

Scientists at Wilfrid Laurier University in Waterloo, Ontario, brought the seemingly distant climate issue home for thousands of Canadians when they created RinkWatch, a website that tracks outdoor skating conditions.[22] After news media first reported on the initiative, the number of frozen puddles, ponds, and backyard arenas involved jumped from 50 to 425 in a week. Robert McLeman, who teaches geography and environmental studies at the university, told CBC News, "We can start to track what's going on with skating conditions across the continent and then by default track what's going on with winter climate trends."[23]

RinkWatch is a reminder that science doesn't need to be conducted in labs by professional scientists alone. "What we really need is public involvement, public engagement," argues McLeman. This kind of citizen science involves getting a group together, one that has a shared purpose that members are passionate about, like watching ice rinks if you love skating or ice hockey. Or monitoring how bird nesting and migrations are influenced by rising temperatures, if you love birds. Some might connect through watching tornadoes or counting monarch butterflies—or even house parties with environmental themes.

Xcel Energy piloted Halloween Gone Green in Colorado, offering customers a kit they could use to host a house party. The kit included games about energy-saving opportunities in the home. The parties were a form of what social psychologist Albert Bandura would call social learning in that hosts tended to be energy-efficiency leaders and the parties allowed them to talk about energy efficiency and show guests what they had done already in their own homes. According to Chris Dierker of Xcel Energy, the parties effectively demonstrated how trusted peers were taking actions to save energy.[24]

Most of our current homes and public buildings have huge potential for better energy use. We can't really blame the energy wasted in buildings and transportation on bad motives, laziness, or vice. A lot of waste is due to history and habit, to a past when energy use was not seen as a problem. And in the present, people have kids and day jobs; they have to hit their marks. To use less energy means considering new ways to conduct life and business. And old habits are often wasteful. Yet we're too busy—focused myopically on immediate needs—to notice the waste. People, and organizations, take the course of least resistance. Realizing energy savings is something that is better done together with your neighbors, a community effort, as emerging solar panel buying clubs and co-ops are showing.[25]

Other efforts focus on public actions like Earth Hour, one hour a year when all businesses and households are asked to shut the lights off and gather around candles. Suddenly the Eiffel Tower in Paris, Tower Bridge in London, and Dubai tower all go dark. Each of us can join in this symbolic act. We do something together to signal a stance, an attitude. Online trolls and highbrow critics are fond of pointing out that this "heroic" action barely makes a dent in the power consumption that day, and that it is meaningless in dealing with emissions. But considering the need for visible social action that helps people bond and makes the issue feel more personal, Earth Hour is exactly the kind of events we need more of.

Hundreds of millions of people and seven thousand cities across the world participated in the action in 2014. To what extent this translates into stronger citizen pressure for ambitious policies is of course impossible to measure directly. But based on social psychological research on modeling and imitation, we can expect it to reduce the cognitive dissonance barrier, and that in turn helps build momentum for change. We and I are doing *something*, and seeing photos and news about what we did.

To use the power of groups for personal change, psychologist Rosemary Randall started a project she called Carbon Conversations.[26] Now a national effort in UK, Carbon Conversations, brings people together in groups to address the reality of climate change. The goal is to reduce their personal impacts, and help one another face rather than avoid the complex emotional issues that arise as people begin to understand the depths of their personal connection to the problem. Carbon Conversations is usually run by volunteers in community groups.[27] A study at Southampton University showed that, on average, individuals reduced their carbon footprints by three tons per year through their participation.[28]

In other efforts, people have banded together for deep-rooted change, aiming to build sustainable, postcarbon lifestyles and resilient communities that can both withstand and alleviate carbon disruption. The Transition Town Movement, started in Totnes, UK, has more than 460 official active initiatives around the world and 600 more getting under way. Transition Town groups can be found in rural communities, suburbs, and cities. Members are encouraged to find ways to reduce energy use as well as their reliance on long supply chains that are totally dependent on fossil fuels for essential items. Local food production is a key. Transition Towns often have community gardens, business-to-business waste exchanges, upcycling programs for reusing old items, renewable-energy initiatives, and an emphasis on building strong local economies, sometimes even through local currencies such as the Totnes Pound. Members gather for meetings, action, workshops, and parties. In the United States there were more than 150 initiatives in thirty-seven states in 2014.

Greensburg, a small city in Kansas, was totally devastated by an extreme tornado in May 2007. Already the following week a few town officials brought up the idea of rebuilding it as a "green" community. After discussion among the fifteen hundred homeless and evacuated residents, most agreed, not least since they recognized that their settler ancestors had been entirely self-sufficient with renewables. They now have the most energy-efficient buildings (LEED-certified) per head not just in the United States but in the whole world. This initiative has put Greensburg back on the map and "is providing an example for rejuvenating rural America by reducing its environmental footprint while keeping citizens safer from severe weather."[29]

Here are yet three more examples of using the power of social networks. First, we might tap into geographic local patriotism by inspiring Los Angeles to compete in greening efforts with San Francisco, New York with Boston, Bronx with Harlem, London with Paris, Copenhagen with Helsinki, and so on. In Europe, for instance, cities have started to compare and compete with each other on the Green Cities Index, to improve their attractiveness, status, and political clout. Second, we could use social networks to engage special-interest clubs and sports teams (say, skiing or reef diving organizations) in climate disruption awareness. The Green Sports Alliance is doing exactly this. Information and messages are distributed via leaders who are much closer to the target groups than are climate scientists—leaders whom group members already ally themselves with. A third approach could begin with pinpointing those people who identify as the climate Alarmed. These folks—approximately forty million Americans—are spread out across the country, and already participate in a multitude of social groups. Activating them by encouraging them to reach their peers through whatever social networks they participate in could give climate messaging a much broader peer outreach.[30]

All in all, being part of an eco-network is one of the biggest determinants of pro-environmental behavior.[31] It might not be entirely clear whether social networks spread pro-environmental behavior among their members, or whether people with a preexisting habit join these sorts of social networks. But what we do know about "social networks and the diffusion of behavior in general suggests that sustainable behaviors will be enhanced by targeting social networks rather than individuals," concludes Adam Corner, a climate psychologist.[32]

Just how many nongovernmental organizations are out there working to address environmental and social justice issues? The sheer numbers of these groups was largely unknown until activist and writer Paul Hawken documented their proliferation in his book *Blessed Unrest*:

> I initially estimated a total of 30,000 environmental organizations around the globe; when I added social justice and indigenous peoples' rights organizations, the number exceeded 100,000. I then researched to see if there had ever been any equals to this movement in scale or scope, but I couldn't find anything, past or present. The more I probed, the more I unearthed, and the numbers continued to climb, as I discovered lists, indexes, and small databases specific to certain sectors or geographic areas. In trying to pick up a stone, I found the exposed tip of a much larger geological formation. I soon realized that my initial estimate of 100,000 organizations was off by at least a factor of ten, and I now believe there are over one—and maybe even two—million organizations working toward ecological sustainability and social justice.[33]

Let me repeat: One to two *million* organizations working for the bene-fit of the disadvantaged and the more-than-human world, if Hawken's and his team's work are correct. That is an astonishing number, a testimony to the social engagement of two-legged apes, who have not completely forgotten their debt toward and embeddedness in the world. If you look at the science that describes what is happening on earth today and aren't pessimistic, says Hawken, "then you don't have the correct data. If you meet the people in this unnamed movement and aren't optimistic, you haven't got a heart. What I see are ordinary and some not-so-ordinary individuals willing to confront despair, power, and incalculable odds in an attempt to restore some semblance of grace, justice, and beauty to this world."[34]

Hawken argues that these groups form the largest social movement in the history of earth. Even if it is mostly invisible to the media, and doesn't work in a controlled and coordinated way, its members are linked by their values. The movement has no manifesto or doctrine, no overriding author-ity. Still, even if it is not united by the one big idea, it offers in its place thousands of practical and useful ones.

## I Party Therefore I Am

A fundamental principle of activism is that none of us can change the world alone; the task is always way too big. None of us can even change our neighborhood alone. Or town. Therefore, we do as much as we can, while acting together with others.

This reality explains why we need to make climate communications as social, interactive, and local as possible. Living creatively with climate change happens when the dispassionate statistics are relinked with social meaning particular to place.[35] Thus, New Yorkers make sense of global warming when hearing how the city's subway system will flood in storms like Sandy. New Orleans residents begin to understand sea level rise when learning how much of their shoreline vanishes each year. Californians grasp the meaning of extended droughts when hearing about their own shrinking groundwater supplies and what that will mean for their daily habits, like bathing, washing dishes, and watering lawns and gardens.

Don't describe a vision of a sustainable Dallas when you're in Detroit—or vice versa. Speak of people, places, and spaces where you are. Refer to local species, jobs, and traditions. Make it transparent, by using the social web to extend social influence and interaction (as, for example, through Avaaz, the global online activist network whose clicktavism spans issues from climate to human rights).

By employing social norms through imitation, status, cooperation, and competition. By involving a multitude of groups in creating shared meaning, we can make climate communications less universal and more social. When we do, we circumvent the first major barrier to conventional climate communications: psychological distance in time, space, and locus of control. The prominent message should not relate to the Arctic ice by 2050 or Antarctica by 2100, but to practical interaction with people and places I care about today. My voice reaches farther through the groups I participate in, and the distance between me and others with the perceived capacity to act is reduced. My sense of in-group can eventually become extended.

This strategy can also reduce the political and cultural polarization at national levels: *Never mind the fighting in Congress, but here in Louisiana, we need to build our own preparedness and resilience.* Or as Kahan writes: "The influences that trigger cultural cognition when climate change is

## Social: Use the Power of Social Networks

Use social norms to motivate others to:
- Reduce power and water consumption.
- Spread social norms through green products and services (rooftop solar, eco-apps).
- Improve recycling efforts.

Use groups and word of mouth from trusted peer messengers to:
- Clarify the scientific consensus.
- Join Earth Hour and similar initiatives.
- Set up home parties; solar panel buying clubs; local-patriotism climate conversations.
- Introduce the topic of climate in existing networks (churches, clubs, sports, and the like).
- Join Carbon Conversations and Transition Town efforts.

Set up funding for social network climate initiatives.

addressed at the national level are much weaker at the local one. When they are considering adaptation, citizens engage the issue of climate change not as members of warring cultural factions but as property owners, resource consumers, insurance policy holders, and tax payers—identities they all share."[36]

The purpose of this networking strategy is not only the direct reduction of wasteful consumption of energy or water in absolute numbers. Such a reduction would of course be welcome and significant[37]—but still more important is the spillover effect into greener attitudes, norms, values, and ultimately votes.[38]

At the same time, remember that groups hostile to climate action use similar dynamics in the opposite direction. They employ their group norms to reinforce the status quo and promote inertia.[39] To reach into contrarian groups, we have to combine this social strategy with others, such as new ways of framing, nudging, and storytelling. Ridicule and sarcasm against an out-group mainly evoke further resistance and polarization, an us-versus them situation, which denialism thrives on.

The more people see happy others conserve energy, install solar rooftops, recycle, shop green, and drive electric cars, the more they are inclined to support ambitious climate policies on local, state, and national levels. By seeing solutions in action, their feelings of helplessness and of being overwhelmed by global climate facts are eased. By seeing and believing that others—neighbors and friends—are taking action on the climate message, they start perceiving it as more personal, nearer, and more urgent, too, counteracting the barriers. Demand for clearer political leadership grows from the bottom up: "We're doing something about it; why aren't the politicians?"

# TEN

# Reframing the Climate Messages

With climate frames like these, who needs enemies?

- **Disaster.** "Experts warn of climate mayhem . . . more frequent and intense heatwaves, bushfires, floods, drought and landslides."[1]
- **Destruction.** John Kerry called climate change a "weapon of mass destruction"—the "world's most fearsome."[2]
- **Uncertainty.** "There is a great amount of uncertainty about climate science. These uncertainties undermine our ability to determine how $CO_2$ has affected the climate in the past."[3]
- **Costs.** "Policymakers [move] toward costly regulations and policies that will harm hardworking American families and do little to decrease global carbon emissions."[4]
- **High Price.** "And the question then is what do we do about it and how much it will cost the consumer?"[5]
- **Loss and Sacrifice.** "People 'must be willing to make sacrifices to cut climate change.'"[6]

For decades now, climate change has been framed[7] as disaster, destruction, cost, uncertainty, and sacrifice. A broad examination of media reports from six countries showed that the two dominant framings were disaster and uncertainty. This was true for most of the climate change news stories circulated to at least fifteen million, across the different media and in different political contexts.

The disaster frame appeared in more than 80 percent of news stories. Uncertainty was the second most common frame, present in nearly 80 percent. Opportunity frames were the third most common (27 percent), but stories using them were mostly about the opportunity of doing nothing about climate, such as fewer cold spells and friendlier farming conditions in the Northern Hemisphere. Only 2 percent of the news discussed positive opportunities such as better energy efficiency, or the growth of solar and wind power.[8] If nothing else, this confirms that journalists and editors are really attracted to doom and gloom.

But as we saw in chapter 3, such negative frames rarely work to motivate people, and have so far certainly not shifted the public to support more ambitious policies. They may be good for selling news and magazines, but when overly negative frames are used they tend to boomerang, spawning gloomy emotions and causing people to avoid the topic.

Fear and loss don't sell. Uncertainty kills determination. Let's therefore try shifting toward frames that support the issue rather than backfire. New frames are now emerging that are more conducive to action. We can begin to talk about climate in terms of insurance, health, security, preparedness, and, most of all, opportunity.[9]

## From Cost to Insurance

The psychological research reviewed in part 1 shows that people are generally loss-averse. They hate rising costs. They hate losing money, jobs, or other stuff about twice as much as they enjoy gaining the equivalent. But, ignorant of this effect, many environmentalists have in combination with the media done a terrific job of depicting the problem exclusively in terms of loss. We're told we'll lose beautiful forests, butterflies, birds and streams, and even human dwellings, coral reefs, polar bears, snow and ice. Worse, the climate solutions have also been framed as losses to us: We're going to lose the possibility to travel where we want, eat meat, or shop freely.

Environmental economists, too, have been prone to use the cost frame in many ways. First, the polluter should pay the true cost. Second, their preferred tool for analyzing policies has been the so-called cost-benefit analysis. Third, a global tax should be put on carbon emissions, raising the consumer costs. And then, fourth, in the absence of a global treaty with a price on carbon, they've argued against subsidies and regulations from a "too-expensive" frame: It's not cost-effective compared with the ideal solution, which is a global carbon tax. In sum, the unintended effects of the cost framing swallow up the good logic of the underlying economics.

People and politicians are repelled rather than attracted to their cost- and tax-speak. It hasn't flown politically.

Rather than going on and on about losses and costs today compared with the future costs of global warming, it is more productive to reframe climate in terms of risk and insurance: How do we insure ourselves today against further climate disruptions tomorrow?

Modern countries do not keep a military defense because it is profitable or has a low cost. Quite the opposite. We do not believe that there will be a military invasion soon. Still, the logic springs from the recognition that large wars *could* happen again. One way to utilize this framing for climate could be: We must build a climate defense today so we can avoid the climate declaring war on us in the future. Future climate mayhem may release waves of unrest, riots, refugees, and terrorists. Good climate policy is good military strategy—something the US Department of Defense has emphasized in its own reports.[10]

We also buy fire insurance even though we're not convinced that our house will burn down later this year or the next. Yet most households and businesses spend a lot of money to insure against such risks. Fires do happen. And when they do, the impacts are severe.

Within this new insurance framing, the climate discussion will turn to questions such as: How much is it worth to pay today to avoid a burning planet in the future? How much more important is it today to insure against military attacks than against climate change? Around the world, we pay taxes to maintain armies to protect the values we believe in at the level of 2 to 3 percent of global GDP. We also pay 3.5 percent of the global GDP per year to the insurance industry against risks such as theft and fire.[11]

The same could be the case for climate action: Even if the probability of large, irreversible consequences is low, it still makes perfect prudent sense to pay a little insurance today to avoid huge future disruptions. How much would this insurance cost?[12] If we are prepared to pay the equivalent of 3.5 percent of total annual output to guard against fire and theft, then why not pay even a 1 percent premium to protect against catastrophic climate disruptions?[13]

In fact, the insurance industries have already started factoring the risk of more frequent severe weather events into their calculations. After seeing their climate-related payouts increasing, many senior insurance executives have been on the forefront of championing ambitious climate measures. It can be no coincidence that the insurance industry, parts of the investment community, and the military are at the forefront of discussing, or planning for, the risks from climate change. This is because they are used to dealing with risk every day.[14] Such numbers have meaning when communicating climate in an insurance framing.

Many mainstream scientists and economists like the UK's Lord Nicholas Stern argue that climate change risks are greater in size and probability than anything we normally insure against. But some of the risk assessments that ordinary people make are on the same time scale as climate impacts—for example, taking out a pension policy that will pay you in forty years. This is the same sort of period over which we expect to see some major impacts from climate change.[15]

A project named Risky Business has applied this framing to the US economy. Its leaders—former New York City mayor Michael Bloomberg; former US Treasury secretary Henry Paulson, who served under George W. Bush; and billionaire investor Tom Steyer—draw a comparison between upcoming climate disruptions and the 2008 financial collapse. Both are examples of the need for good risk management. The Great Recession, they say, was a warning of what happens when society fails at that management. In the words of Paulson, also a former CEO of Goldman Sachs: "I know a lot about financial risks—in fact, I spent nearly my whole career managing risks and dealing with financial crisis. Today I see another type of crisis looming: A climate crisis. And while not financial in nature, it threatens our economy just the same."[16]

The report released in 2014 by Risky Business argues that investors have largely been kept in the dark about how climate change will impact specific industries or specific regions. It explains that acting now could spare American businesses and the American economy from the risk of becoming extremely vulnerable to future losses.[17] They frame climate change as "a crisis we can't afford to ignore,"[18] which is turning the conventional cost framing on its head to make it work for the better.

Risky Business hopes to motivate change on a national scale. Another documented approach is to frame today's climate policy expenditure as just a little bit smaller *increase* in future income: National incomes keep growing; we can afford to shave off a tiny bit of that extra gain to insure ourselves better. This approach is called not losses but "foregone gains."[19]

This reframing can work at the individual level, too—tapping into people's desire to avoid future losses rather than realize future gains. For instance, when communicators talk to homeowners, they could frame the need to buy new, efficient appliances as helping the homeowners cut energy bills immediately, instead of helping them save extra money years in the future.

## From Destruction to Health and Heart

Pictures of homes with smashed windows and broken-off roofs, waves breaking through harbors, broken barracks and coastal buildings are becoming all too familiar. But such climate iconography, even if evoking a dark fascination with the violent, does little to inspire lasting engagement.

A more promising frame views climate action through the lens of improving health and our quality of life. Universal framings such as *global climate*, *the planet*, and *the environment* inadvertently strengthen the distance to the topic. It is all thousands of miles out there and far away in abstract space, measured in averages on graphs. What makes lives worth living, however, is more about people, our love of family, hometown, friends, and children. And their health in particular.[20] So we need to tap into the tacit, deeper frame that everyone knows but that often slips out of mind: that the health of our human lives depends on the vitality of the more-than-human world. Healthy and vibrant human lives always happen inside specific locations, landscapes. And these places each have their set of winds, temperatures, trees, streams, scrub, squirrels, worms, crickets, and all the other singing, fluttering, and feathered creatures. Together they form a closely interlinked culture–nature nexus.

Global warming will force changes in both our cultural and our wild landscapes. A core part of our identity may then disappear. Both conservatives and progressives care about natural landscapes and their beauty. But with a lot more severe floods, droughts, and heat waves, roads that fail as well as homes destroyed, lives and health will be put at risk.

There will be more itching, sneezing, swelling, and gasping for breath as the climate shifts, expanding the ranges of poison ivy, mosquitoes and other stinging insects, pollen-producing plants, and Lyme-disease-bearing ticks. One Pennsylvania report explains there'll be increased asthma, respiratory disease, and heat-related deaths.[21]

The new idea is that the most effective way around climate-policy ambivalence is to invoke imminent dangers to human health: "'What's killing me today?' with emphasis on killing and me and today."[22] With more $CO_2$ in the air, plants produce more pollen; pollen counts are projected to double by 2040. This is no good at all for allergies and asthma. The World Health Organization is warning that air pollution is responsible for one out of every eight human deaths, largely because combustion of fossil fuels results in invisible airborne particles that get lodged in our lungs and suspended in our blood.

On the solutions side, and still in the health frame, it could be excellent for both our health and the climate if we were to eat less meat raised in fossil-fuel-intensive settings and more sun-based, short-traveled fruit, nuts, and crunchy vegetables. It is good for both the body and the climate to bike more and drive less. Where the air is fresh and unpolluted, we breathe better. Spending time outdoors, in a landscape where plants and animals have temperatures and rainfall to follow their own boisterous flows, seems to vitalize and inspire human beings. There are other tangible benefits, too: Trees seem to be mitigating some effects of air pollution in the United States, saving around seven billion dollars a year in human health costs. The three causes—human health, vibrant landscapes, and a stable climate—have much in common.

Doctors have identified six categories of health impacts: heat-related sickness, respiratory health problems, infectious disease, waterborne disease, food insecurity, and mental health problems.[23] Some of these health implications of climate change are relatively well understood (for example, an increased likelihood of heatstroke and violence), while others are less obvious (such as the rapidly rising rates of asthma and respiratory conditions). Drawing awareness to the human health impacts seems to be an effective method for elevating public concern in the United States, helping to frame climate disruption as a concrete, personal concern for everyone.[24]

People's emotional reactions to climate messages have often been overlooked both in research and in communication efforts. Virginia-based researcher Teresa Myers and colleagues did experiments that focused more on the affective component of climate engagement. Participants were asked to indicate which parts of the framed message made them feel hopeful or feel angry by clicking on those sentences. One such health-frame sentence was:

Redesigning our cities and towns to make it easier and safer to travel by foot, bicycle and public transportation will reduce the number of cars on the road, reduce carbon dioxide emissions, reduce traffic injuries and fatalities, and help people become more physically active, lose weight, strengthen their bones, and possibly even to remain mentally sharp as they age.

Fifty-seven percent of subjects responded with hope to this sentence. On the other hand, the sentences that framed climate inside a security frame elicited more angry responses overall:

The most recent Quadrennial Defense Review—a national security report prepared every four years by the Pentagon for the US Congress—concludes that global warming is a "key issue" likely to harm US national security in many ways. They also argue that efforts to limit global warming are a "win-win" because they will reduce the risks of global warming and improve America's national security.[25]

Possibly, the Doubtful and Dismissive segments perceived the text to be an attempt to force a link between an issue they may care deeply about (national security) and an issue that they tend to dismiss (climate change). Or they felt the statements were attempting to co-opt values they care strongly about, thereby producing the angry reaction. Myers's research is thus a much-needed contribution to testing out how these frames are actually received by different cultural segments.

## From Uncertainty to Preparedness and Ethics

For scientists it's easy to fall into the uncertainty frame, since climate change is about long-term impacts on rainfall, glaciers, storms, and so on. Scientists are experts on uncertainty and probabilities, and can go on and on about technical details on probability calculations and error bars for hours. What the public hears, however, is that "They don't know!" Denialists have grabbed this gift, this opportunity, and inflated it as a weapon against statements that climate change is human-caused, and that it will get more destructive in the near future. "Since it's so uncertain, we shouldn't do anything," goes the very predictable refrain. Some climate activists or alarmists have then been counter-arguing by insisting that impacts are certain. But unwittingly, by negating the uncertainty, they actually activate and *reinforce* the uncertainty frame.

Rather than arguing about uncertainty, there are two new frames that can replace it altogether: *preparedness* and *ethics*.

Preparedness has to do with how to get ready for upcoming change, strengthening our resilience and feeling of safety. No matter what comes, we'll be ready. In essence it says: Yes, extreme shifts may be unpredictable, but they're far from unthinkable. Therefore, we are better safe than sorry. We bring a raincoat for the mountain trip, despite the fact that the sun is shining. We put on sunscreen before hitting the beach. We put a seat belt on before driving. We brake before the bend, not because the car can't make the turn at our current speed, but because it's safer and more comfortable at lower speeds. And now we'd better prepare for the upcoming "bends" of higher floods, longer droughts, wilder storms, wildfires, and more.

In this frame, getting ready is prudent and common sense, while inaction is reckless, unethical, and irresponsible. Simply adapting after the accident, wildfire, or flood is too late; proactive preparedness is needed.

A focus group study conducted in the United States by the foundation ecoAmerica found that 85 percent of voters seek preparedness as the preferred approach to address climate disruption, whereas adaptation is less effective, falling 16 percent below preparedness. To mainstream Americans adaptation is a passive, reactive strategy.[26] Studies of psychological preparedness in the disaster and public health arenas show the importance of proactive responses. Being prepared, building resilience, and keeping a weather eye on potentially serious future threats simply makes good sense and builds a sense of coping and mastery in face of unwieldy threats.[27]

After decades of experience, scenario planner Arie de Geus said that the only relevant discussion about the future is one where we "succeed from shifting the question from whether something will happen to *what we would do if* it did happen." Part of the *preparedness* frame—as it deepens in our understanding—includes slowing down the speed of warming and reducing risk by reducing our emissions. The ecoAmerica study also concluded that talking about *preventing pollution* and *protecting the air we breathe* persuades many more people than conventional uncertainty frames. When we start preparing for it, climate change becomes more real, near, and urgent, and willingness to mitigate can go up, too. By taking action to prepare, we're nudged into seeing the issue as more politically important, a point I'll return to in the next chapter.

The other frame to replace the worn uncertainty frame is the ethics frame. Anti-climate pundits often argue: "Since other countries, like China and India, are bigger emitters and won't curb their emissions, increasing our own costs to decrease our emissions is a completely wasted effort." A variant of this goes, "If the climate is changing, the impacts of cutting emissions now are highly uncertain. High cost—no gain."

But climate disruption policy is really about ethics and values, not the uncertainty of the science or other countries' future actions. Laying our values openly on the table will not end the debate but will take the issue of scientific uncertainty out of it. For too long, we have let the uncertainty debate divert attention from the values-based questions that should be guiding climate policy: How far shall we let economic resource consumption grow at the expense of marginalized people, ecologies, and the more-than-human world? How should long-term considerations for future generations weigh up against short-term costs? How to balance state regulations and emission trading in limiting carbon pollution?

Many object to climate science, not because of the science itself, but because they don't like the *policy* implications that have been promoted. But climate science cannot answer such value-laden policy questions. Modeling, measurement, and probabilities used in climate science cannot resolve value conflicts. Conflict resolution, negotiation, and reframing are methods much better equipped for that.

The deep framing of ecological ethics conveys the realization that we're embedded inside the more-than-human world. Destroying the web of life around us implies destroying our relatives and ourselves, even if we may not notice it immediately. Therefore, we ought to curb emissions, not because other countries or other people are doing it, nor for our own benefit only, nor because every predicted climate change impact is 100 percent certain—but because changing is in service of life—protecting and compassionately caring for ourselves, current and future generations, and the other beings we share the planet with.

Change strategist Tom Crompton, who has studied what motivates people to pressure their elected leaders for change, says that shifting to a values-based frame for campaigning makes sense because it supports larger-than-self cultural values, building a sense of common cause.[28]

The philosopher Immanuel Kant would say it is an ethical duty according to universal moral law.[29]

## From Sacrifice to Opportunity

A final and vital framing is the positive *opportunity* framing. If I feel that other people—the environmentalists or alarmists—are forcing me to sacrifice my SUV and T-bone barbecue for the sake of some green future I can't envision, this makes me resentful. If, however, climate change action can promote a better society where people are warmer and more considerate, and if my new electric car has greater acceleration than the old car and the barbecued fish and vegetables contribute along with biking to making my abs and legs look sexier, then we're talking. These opportunity framings can reach audiences that insurance, health, or ethics leaves cold.

The *opportunity* frame emphasizes that buildings, companies, cities, and societies with low emissions are more efficient and competitive while also providing better jobs. To demonstrate industrial and social leadership in these areas is an investment in future profitability and competitiveness. We get better growth *and* better climate.[30] Climate efforts can promote scientific and economic progress, and can make us more caring and compassionate people. These framings would not create the type of backlash that *doom* and *cost* framings do.[31]

This new framing turns the economic development frame in favor of action, recasting climate change as an opportunity to grow the economy in a new, smarter direction. At the Breakthrough Institute in California, Ted Nordhaus and Michael Schellenberger advocate provocatively for a move away from what they call the pollution paradigm—the doom-inspiring statements about dire consequences facing us if emissions are not radically curbed. They argue that only by building diverse coalitions in support of innovative energy technology and sustainable economic prosperity can meaningful action on climate change be achieved. With this framing strategy, they seek to also engage conservatives, who think predominantly in terms of market opportunities, and labor advocates, who value the possibility of job growth.[32]

For example, the conservative US tea party movement is starting to embrace solar energy. Maybe they are getting a taste of a little green tea? This is probably not because they are worried about climate change, but because it creates more American jobs while at the same time freeing customers from bondage to their large monopolistic power utility. People can break free from the grid and become more energy-independent. If you install solar on your roof and battery storage in the basement, you can reduce bills, cut the taxes paid on grid usage, and be freer in your choice of energy supplier and consumption. The new and emerging label for this framing is *free-market energy*. The degree of choice in the energy source is widely appealing to conservatives, particularly as the cost of solar power has declined so substantially in recent years. According to SunVest's Matt Neumann, this rings particularly true in a conservative state like Wisconsin: "This is a national defense issue, a free market competition discussion, property rights, economic development—this is a Republican concept through and through."[33]

For these audiences, one shouldn't preach about global warming, environmental disaster, or ethics for saving the planet. Rather, one should keep spirits up by giving people the choice of free-market energy. And cut the red tape and installation costs for solar and wind, so that households that want to cut their losses on rising energy bills can easily install insulation and solar power when and where they want to.

Clean energy is no longer alternative; it is attracting a serious amount of money and entering the mainstream. Bloomberg New Energy Finance analyst Ethan Zindler explains:

> I think there's definitely a libertarian flank on all this, which is people like the idea that they are producing their own energy and using it themselves. And they don't like the idea necessarily that they have to pay some kind of fee for this opportunity. We've seen some of that particularly in Georgia, where regulators have tried to crimp on solar development . . . and also in Arizona, where one of the group's really supportive of the solar industry is run by the son of Barry Goldwater, [who was] a quite conservative presidential candidate.[34]

The frame can also be turned around from *regulations* to *market*: Are you a supporter of the free market? Then don't tax the sun. Advocate instead for letting the free market solve the energy and pollution problem by putting a price on fossil emissions, as in British Columbia.

British Columbia already has a revenue-neutral carbon tax. This returns the funds of carbon taxes directly to the citizens by decreasing people's income taxes. The system enjoys broad support from 64 percent of citizens.[35] The province's economy is doing well and its greenhouse gas emissions are falling. They've shown that a well-crafted climate solution can work. More income, lower emissions. The debate should be about how to best achieve maximum economic benefit while reducing greenhouse gas emissions.[36] Steadily increasing emission costs on centralized power plants would spur the already rapid growth of distributed free-market energy.

Clean energy is also now creating more new jobs than the fossil fuel industry. The US solar industry alone already employs more people than coal and gas industries combined, and is growing vigorously.[37] Renewable energy and energy efficiency create more jobs per unit of energy than coal and natural gas. Ambitious efficiency measures combined with 30 percent renewable portfolio standards can generate over four million full-time-equivalent job-years by 2030.[38] Thus, green buildings are a win–win–win proposal, good for jobs, economic opportunity, and climate. "Greater building efficiency can meet 85 percent of future demand for energy in the US and a commitment to green building has the potential to generate 2.5 million jobs," writes the US Green Building Council.[39]

## We're in It for the Long Haul

All the new framings introduced here are about people and their lives, lungs, safety, values, and—not least—local job opportunities. This moves climate away from being a separate, low-priority issue and integrates it into the top policy concerns of citizens: economy, jobs, and health care.

When communicating climate we should never accept the backfiring frames (doom, uncertainty, cost, sacrifice). Don't negate them, or repeat them, or structure your arguments to counter them. That just activates those frames, thereby strengthening them in the audience's mind. Always go on the offense with your framing, never defense, states framing expert George Lakoff.[40] In communicating climate change we want to train, repeat, and shift the balance toward the supporting frames: that more commercial and

## Supportive: Use Positive Framings

When speaking of climate, frame it as:

- Insurance against risk.
- Health and well-being.
- Preparedness and resilience.
- Values and a common cause.
- Opportunities for innovation and job growth.

political action is needed right away to ensure safety for society, secure our health, be prepared for what comes, and realize the amazing job opportunities that the shifts in clean energy are bringing.[41]

But we must also remember that simply using some new words or better slogans in the coming months is far from sufficient to shift the public's understanding. As Lakoff points out, the long-term climate issue needs a much stronger communication system. In addition to serious framing research institutes, such a system needs training facilities, spokespeople in every electoral district that work through social networks, and publicists to get through to the media.[42] Otherwise the news journalists and editors will—if they dare mention the C-word at all—keep harping on the old doom and uncertainty frames. The issue is much too important to be stuck in an apocalypse-movie mode!

None of this, though, means we only have to focus on today's battles and forget the larger, long-term picture. In addition to using frames that prompt near-term opportunities and policies, we also need to develop the deep frames that are needed in the long run. The effectiveness of short-term framing depends on how deeply embedded the long-term frames have become. And the success of both depends, as Lakoff concludes, on how well we communicate the systemic nature of climate change's causes and solutions.

Finally, framing is about much more than language. It is about institutionalizing these frames in regulations, law, market design, NGOs, and think tanks. We need to think constantly about what framing gaps exist and how we can fill them. How can the supportive frames get institution-alized? How can better understandings of framing guide policy making? This represents nothing less than a culture change, albeit a gradual one.

Framing is inescapable in all human communication. To be visible at all, truth must be framed somehow. That is why it is so critical to build supportive long-term climate frames.

# ELEVEN

# Make It Simple to Choose Right

If people lack green alternatives to heat or cool their houses, or get to work in a practical manner, then efforts to create long-term engagement in climate will eventually fail, no matter how well we utilize social networks and framings. Dissonance will prevail and apathy will spread. More information about how critical climate change is or more facts about the benefits of energy-efficient insulation or appliances will lead to little change, if there are no convenient and simple ways to get them home and correctly installed. When life is crammed, time-demanding to-dos slip downward on our priority lists. We've other stuff to do than to think energy efficiency and climate.

What we choose to purchase depends not only on price and technical information but even more on *how* the choice is presented. There is huge potential in making the right choice the default option. Consider organ donations, in which defaults may save lives. Some countries, like Germany, have an opt-in policy (you have to register to become an organ donor) and others, like Austria, have a default-based opt-out policy (you're a donor unless you register not to be). In Austria, 99 percent of the people are organ donors whereas in Germany the consent rate is 12 percent. Such differences are not due to Austrians being much more altruistic and kind. The fact is, most people simply do not bother to change the default on their donor forms.[1]

Similar effects have already been demonstrated for human behavior in areas such as health, pensions, finance, law, and marketing. For climate and nature, better use of insights from behavioral research can translate high concern into visible and pervasive actions. This approach, called nudging, builds on research from psychology and behavioral economics. Small changes in choice architecture may have a large impact on consumer behavior, sometimes even larger than that of price.[2]

To avoid dissonance and maintain engagement, as many of our daily actions as possible should be consistent with climate knowledge. At the same time, the green actions should not demand too much extra effort—the breaking of entrenched brown habits from the twentieth century is a demanding task.[3] The long-term value of engaging people at the personal level is that it builds public support. In the absence of political leadership, each person and each choice becomes more important. We need to reach a critical mass of both before businesses and policy makers can make the great swerve happen.

## Green Defaults for Paper, Planes, and Parking

As we saw with organ donation, the simplest nudging approach involves the use of default settings, which apply when people make no active choice at all.

Take paper. One example of a green nudge is the zero-cost option of simply changing the default on all printers from one-sided to two-sided printing. Rutgers University made double-sided the default setting for all its student printers, and its New Brunswick campus saved over eighty-nine million sheets of paper during the first six years of the conservation program.[4]

In a broader study, the double-sided default reduced paper consumption by more than 15 percent overall, and the effect remained intact over a period of at least six months. If applied to all US offices, this would offset the carbon emissions of roughly 150,000 cars per year.[5] It seems we can often save money and slash emissions by asking people to do nothing.

Or take planes. Another climate nudge might involve automatically including your share of the flight's $CO_2$ emission allowances when you buy a plane ticket—not as a tax, but as a default charge.[6] You would actively have to opt out in order not to pay, preferably far down on the fifth or so page of the web form. The airlines could retain a certain cut of $CO_2$ allowance sales, so they would have an incentive to develop the system.

Or take city parking. We could make it faster, easier, and more comfortable to take public transportation than to drive and park by improving public transportation—lowering prices and increasing frequency, for instance—while at the same time restricting parking in city centers. Such large-scale nudges have transformed the traffic and congestion in the cities of Curitiba and Bogotá,[7] as well as inspiring Los Angeles to start making changes.

## Green Nudges for the Home

Many of the choices that consumers make have large and long-lasting consequences for energy consumption and therefore emissions. This is particularly true when we buy houses, insulation, cars, household appliances, clothing, and certain foods, especially red meats. An inefficient air conditioner, freezer, or dryer will continue to waste power over its entire lifetime, maybe decades from now. Can we modify choice situations to help make climate-friendly purchases easier?

The Norwegian nonprofit foundation GreeNudge collaborated with Elkjop, a retailer of electrical products, to get people to buy energy-efficient dryers by providing life-cycle costs that include power usage in large fonts next to the purchase price. As consumers make their decision about what to buy, they can clearly see the benefits of investing in the most efficient appliances. If this nudge was applied to all purchasing situations to favor the most energy-efficient electrical appliances, it could lower average appliance power consumption by up to 5 percent. Is this much? Yes, when you consider that it is achieved simply by rearranging sales labels in the shop, a nearly no-cost option that actually gives the shop extra revenue from higher-quality dryers being sold.[8] And the results, if applied to all appliances in the EU alone, would be the carbon equivalent of taking around two million cars off the road.[9]

In the UK, around 40 percent of all households have insufficiently insulated attics. This leads to high energy consumption and bills. Extra insulation makes a lot of sense for both homeowners and society. But what does it take to insulate your house? Well, first of all you have to clean up your attic. All the old tennis rackets, photo albums, furniture, and memorabilia have to be taken somewhere else if the insulation can be effectively installed. But that's a lot of hassle.

Offering people subsidies for home insulation did little to get people off the couch and cleaning the attic. And offering discounts to motivate households to insulate their attics had next to no effect. But when UK officials simultaneously offered an integrated attic-clearing service—the homeowners did have to pay for it—they saw a fivefold increase in households that agreed to insulate their attics.[10] The nudge here is to bundle two services into a one-step versus a two-step decision.

What about choosing green energy programs from your utility service? In the German town of Schönau, more than 90 percent of people are enrolled in green energy programs. This is a strikingly high level of clean energy use. Here, as well as for 150,000 customers of the southern Germany Energidienst company, green energy programs are the default. Even after customers received their bills and noted a slightly higher cost for green energy, most chose not to opt out of the program—more than 90 percent. This is a dramatic contrast with the relatively low level of participation in clean energy programs in other German cities and towns. The nudge here is to require an active opt-out choice rather than an active opt-in option.[11]

In Fort Collins, Colorado, garbage had previously been placed in ninety-gallon containers and recycling in thirty-five-gallon containers. The city switched things around, giving residents the ninety-gallon containers for recycling and the smaller one for garbage. Residents then started recycling more, filling up the larger containers, and had less garbage to put into the smaller containers.[12]

## Nibble Nudges: Meats and Food Waste

If you select a large plate at a buffet, you tend to fill it all up. And maybe not eat it all. One nudging study showed that simply reducing plate size and providing some social cues reduced the amount of food waste in hotel restaurants by approximately 20 percent. Since food production is a major contributor to climate change and other forms of environmental degradation, this is a significant improvement with little investment. The measures can reduce the amount of food the restaurants need to purchase, and there is no change in measured guest satisfaction, thereby making it likely that profits will increase. Such nudges provide an opportunity for win–win–win: Restaurants, consumers, and the climate all benefit.[13]

In 2009, Columbia University removed trays from its dining halls at John Jay and Ferris Booth Commons to help eliminate food waste. Students were taking too much food on their trays, then throwing away 190 to 450 pounds of it at every meal. The tray-less cafeteria now saves fifty pounds of food waste at each meal and three thousand gallons of water waste each day. Reusable and recyclable eco-containers for take-out food and refillable water bottles are given to students. Other universities that have adopted trayless cafeterias include Georgetown, Johns Hopkins, the University of Virginia, American University, and Virginia Tech.[14]

Reducing meat consumption is an effective way of reducing climate emissions, since mainstream industrialized farming for red meats in particular releases far more methane and $CO_2$ than does other food production. So what about shifting the chef's special to a delicious non-meat option more often? An experiment at a restaurant at the University of Oslo tested the effect of renaming what it had traditionally called just the "vegetarian option"—calling it, for instance, Mexican-style taco—and making it the special dish of the day. Results showed an increase in vegetarian selection among customers eating meat on a daily basis, particularly for customers with otherwise low connection to nature.[15]

Is it possible to couple pizza giveaways with energy savings? The Danish energy company Vestforbraending opened a new pizzeria called Vest Pizza in the northwestern suburbs of Copenhagen. The restaurant's sole purpose was to reward the community's aggregate energy savings. In preparation for the cold season, Vestforbraending sent out eight wintertime heating conservation tips to their customers, then subsequently measured the total reduction in winter heating load. The more heating energy was saved by the community compared with previous years, the more pizzas the restaurant would heat up and give away.

On the restaurant's first night, residents earned 173 free pizzas by their shared energy savings of 1,732 megawatt-hours. Happy people queued up to collect their bounty, all smiling.[16] Although the experiment mainly confirms that offering free pizza is a surefire way to achieve any goal, it also highlights a few approaches that may be helpful in designing other energy conservation campaigns: Empower people with specific seasonal advice, tap into the power of community, and highlight the benefits in a tangible, even edible way.[17]

## Nudging Goes Mobile and Moves into the Smart House

Giants in the information technology industry have long battled over market shares for PCs, web browsers, mobile phones, and tablets. Some analysts predict the next battlefield will be the smart home. Apple has already targeted the smart house with HomeKit, and Google has purchased Nest, which makes smart thermostats. This battle will take decades since fixed installations such as thermostats, solar fixtures, and smart appliances live longer than browsers, mobiles, or tablets. What does this battle for smart-house market share mean for the climate? About 60 to 70 percent of all electricity and 40 percent of all primary energy is consumed by buildings (residential and commercial).[18] A large part of this is today wasted. Achieving smart, green buildings that nudge their inhabitants and regulate themselves in smart cities therefore holds vast opportunities for reducing emissions in the coming decades.

The key is first to make residents more aware of their energy habits, then to integrate technology that learns to efficiently adapt lighting, cooling, and heating to our life rhythms. Smart-home devices regulate temperatures according to the presence and schedule of inhabitants, while interacting with them in real time and in a user-friendly way. They can send a simple reminder when they register that you've left the house: "You forgot to turn down the air conditioner, should I do it for you?" Early studies of smart energy-use meters showed power reduction of at least 20 percent just by displaying real-time costs.[19] This is almost certain to improve as interfaces mature.

Since apps live on mobile devices, they are always close, and can grab your attention in real time. Stanford professor B. J. Fogg is a pioneer in the area of persuasive technology. He says that for new behavior to occur, three components must be present: motivation, ability, and a trigger. For example, if you're trying to get yourself to work out or save power, that might mean having the desire (*motivation* or social norms), having the time and resources to get on a bike or go jogging (*ability*), and setting a calendar event to remind yourself to go (*trigger* or nudge). Motivation * ability * trigger = action. The best health and fitness apps use all these components in what they offer. By integrating such apps with energy use in smart houses, we can expect energy waste to be curbed dramatically.

## Green Nudges as Climate Communication

There are a large number of nudges we could implement to improve the climate and environment. The good news from this field is that the same factors that lead us to make a mindless and polluting choice can often be reversed to help us make a mindless better choice. Behavioral economics offers insights about how to encourage less wasteful behavior without inducing the resistance often associated with regulation, higher costs, and red tape.

Of course, it is also true that further progress in actual emission reductions will require the use of more standard tools, too, including economic incentives and regulation. No one contends that nudging, tweaking, or cajoling people into piecemeal behavioral changes like reusing plastic bags or buying efficient appliances can alone solve our climate problems. The more modest suggestion is that a better understanding of choice architecture and nudging will significantly expand the toolbox for climate communications.[20]

In reality, nudges are also methods of climate communication. They help us get around the five barriers:

- They work around the distance barrier by making the climate issue feel near and relevant to personal behavior.
- They nudge us out of the cost and sacrifice framing that haunts the climate issue and creates the doom barrier.
- They promote behavior that influences attitudes, helping us reduce the dissonance and denial barriers. It is easier to behave consistently with our beliefs when nudged.

Research shows that giving money or time to a cause strengthens our positive attitudes about that cause. So nudges that combine thinking and doing can turn cognitive dissonance around for the good: *If I do all these things—insulate, go solar, recycle—then the cause must be important, and therefore the science behind it right.* This seems to be the way our minds work—more psycho-logical than logical.

Actions, particularly our own and those of significant others, speak louder than facts and words, creating a potential virtuous cycle. Behavior influences attitudes, attitudes grow into public support for greener policies, policies are strengthened then shape behavior toward even more climate-friendly actions, which again strengthens attitudes, and the cycle continues.[21] Consider what happened when Stockholm instituted a rush-hour tariff. Citizens were initially skeptical, but after they all started paying and could see that it worked well to reduce traffic congestion, attitudes turned positive. The same happened after restrictions on in-door smoking came into effect; people resisted at first, but after the introduction, people could

## Simple: Use Green Nudges to Make It Simpler to Act

Some examples:
- Make life-cycle costs salient on all appliance price tags.
- Make smaller plates in restaurant buffets the default.
- Make non-meat special dishes a restaurant's default.
- Make double-sided printing the default.
- Include voluntary $CO_2$ price fees in plane tickets as the default.
- Increase the frequency and speed of buses and biking while reducing car parking and access to city centers.
- Bundle home reinsulation with attic cleaning and renovation.
- Make recycling fun with painted green steps, big-belly bins, and the like.
- Host local free pizza parties as a reward for community energy conservation.

feel the cleaner air, and support for regulations increased. From changed behavior to changed attitude![22]

The idea that change happens in the direction from behavior to belief, versus the opposite, may seem surprising. But it is a quite well-established finding in social psychology. Unfortunately it has so far eluded too many climate communicators who think that information is cheap and easy. But facts alone are ineffectual. Nudging behavior is where the rubber hits the road.

# TWELVE

# Use the Power of Stories to Re-Story Climate

As humans we create meaning in our lives through stories. So it's not surprising that grand narratives have evolved about global warming.

Scientists sometimes lose sight of the fact that they are telling a story when fully immersed in data or equations. But any data presentation, no matter how dry, factual, or objective, weaves a narrative. The implicit narrative that underlies a typical climate science presentation goes like this: "Once upon a time there was climate stability. Then came coal and oil combustion, and the temperature started rising. If the burning continues, climate flips into a hot state." This quickly becomes a story in our minds, complete with beginning, middle turning point, and end. The question, though, is whether this is a constructive narrative when communicating to non-scientists.

I don't think there is just *one* right type of climate story to tell to get people to understand the urgency of the issue and move them to action. Rather, a plurality of stories is needed, each creating meaning and engagement for different groups of people. I do think, however, that one story in particular has been contributing to the stalemate of the climate paradox. Since it is so easy to interpret literally, it has become universal and fundamentalist, and has stalled progress toward societal transformation.

The story is the apocalypse of climate hell, and it has been used often, implicitly, and without reflection in climate communication.[1] That is not surprising, because it is a predominant story in Christian, Western culture: Its roots go back to the last book in the New Testament about the end times, with environmental and climate disasters described in exquisite detail as a form of punishment for sin and decay. "If we continue on these evil ways, we will all burn in hell." And "The end is nigh."

Too often climate messages fall into this well-worn story track, even with no conscious intention at all from the messengers. You might say that the apocalyptic story comes uninvited and spreads like a thick woolly fog around graphs, figures, calculations, and news media articles. Described again and again are heat waves, drought, wildfires, floods, sea level rises, extinction of species, and self-reinforcing feedback mechanisms that may escalate greenhouse gases in the atmosphere. In other words: The end of our world as we know it is coming soon, and it is all due to our sins. It is a powerful archetypal story.

It is not impossible that the future will in fact be something similar to a climate hell, but that is just one story, just one type of scenario—and it's one that generates fear, guilt, anger, despair, and helplessness as its shadow side in the here and now.[2] We're so tired of this story by now that it is a relief when comedian Stephen Colbert makes us laugh at it:

Now Nation, I have spent the last week in a rage over the Obama administration's new 800-page national climate assessment that claims we're ruining the environment. It made me so angry, I printed it five times . . . But then I read the report, and I have to admit, it is so terrifying, that it left a carbon footprint in my pants.[3]

There are abundant defeatist accounts about our shortcomings and inevitable failures. "What we do wrong" is an addictive, repetitive narrative. We need to tell other stories with other imagery and emotions associated with them. "To be truly radical today is to make hope possible, not despair convincing," as Raymond Williams once said.

There are many stories of what is going well, of conviction and endurance, as well as stories that describe and help us imagine a renewal of society, wildlife, and ecosystems. We could tell tales about the people who stand up against destruction and accomplish spectacular feats. There are well-known eco-entrepreneurs like Elon Musk of Tesla, who changed the world of electric cars; Paul Polman of Unilever, who ushered in plans to halve the company's environmental impact by 2020; and Janine Benyus, who spearheaded the biomimicry initiative that looks to nature as a model for sustainable design.

There are also untold stories about people who care and act on the basis of vision, determination, and joy, or who demonstrate resilience after mayhem. We need new stories to make sense of the ongoing boister-ous transition toward the greening of technology, business, and culture. Creativity and capacity for innovation appear in stories about small-scale solutions such as solar cookers, electric bicycles and buses, and soda bottle lightbulbs—plastic liter bottles that are filled with water and installed in roofs to light up rooms. There are stories about bioenergy systems, about upcycling of waste to high-value products, and about passive houses, net-zero houses, and even plus houses, which make more energy than they use, all from renewable sources. Large cities all over the world are joining the competition to transform into green hot spots.

Typically, says climate psychologist Adam Corner, the communication challenge is thought to require documentation of public attitudes, tested models of behavior change, and the rigorous roll-out of social scientific research. But this approach tends to lose the creative *art* of storytelling. "All of this is true," he writes, "but it is human stories, not carbon targets, that capture people's attention."[4]

There are stories to be told of scientists who are achieving wonderful things, discovering magnificent ecological relationships and amazing but vulnerable behavior of animals such as tortoises, leopard seals, or monarch butterflies. We can tell stories of the surprising opportunities arising from smarter relationships between economic production and connected ecologies. Damaged or unproductive land can return to being forests and wetlands, and nature can demonstrate its often marvelous ability to restore vital ecosystems. Many wild species can settle surprisingly close to urban areas, as long as humans do not destroy them. Stories like this would stretch the horizon farther than just working to stop emissions, stop the destruction, stop everything. "Go-stories" would rather describe an ecologically richer, re-wilded, better world that you and I would look forward to living in.[5]

There is no shortage of ingenious solutions for green growth that can be translated into inspiring stories.[6] However, there does seem to be a shortage of captivating storytellers who spread inspiration as well as vivid and attractive images of a future in which we live with more decent jobs, greater well-being, and lower emissions alongside recovering forests. If it cannot be imagined and well told, then people will surely not work for it to happen.

Among the profusion of emerging new narratives we find green growth, technological optimism, health, happiness, women's rights, solidarity, resilience, self-identity, ecological restoration, wonder, and a sense of the sacred.[7] The more we tell these stories, the more we will begin to live them.

## The Green Growth Narrative

The story of brown growth—that only more fossil fuels can secure progress—has long bullied us in public discourse. But the story of green growth can replace it. The core of that green growth story is that we can continue economic value creation measured in money as we simultaneously reduce the physical flow of materials in the economy measured in tons and kilowatt-hours. In a nutshell: More value with less resources wasted. Or put in a more precise way: We can and should grow those types of economic activity that at the same time lead to a smaller ecological footprint.

Why will this happen? Mainly because it is more profitable to be smarter and less wasteful than today. Our current economic system is extremely wasteful seen through a material flow lens. Vast amounts of stone, ore, soil, water, steel, coal, cornstalks, rice husks, timber, and more are moved around at a frantic pace, but less than 1 percent of it all ends up in durable products. And very, very few of those durable products are then recycled or upcycled back for further value creation. This outdated production system was built in previous centuries to maximize profits from labor and capital while ignoring the resource flows.

The new story goes along these lines: At the time of the first industrial revolution, people were few or scarce, while nature was huge and abundant. The way to larger wealth was to specialize the tasks of people along with new machinery, so they could do more per hour. It made perfect sense at the time: Better labor productivity meant more machines.

Imagine you're a visionary in 1750, and you travel (by horse) to the British Parliament. There, you start to give an eloquent description of the future fifty years later, in 1800. In 1750, weavers and spinners had made approximately the same amount of yarn for centuries. To hand-weave a twelve-pound piece of eighteen-penny weft took fourteen days.[8] Now imagine proclaiming that in just fifty years, one person would be able to do the job of two hundred in a day! You'd be ridiculed and thrown out. Or the weavers would smash you for threatening their jobs (which they tried with James Hargreaves, the inventor of the spinning jenny).

The same happened all over again in the second industrial wave, when railroads started to be built in the 1830s. Imagine someone coming to Washington with an exuberant voice claiming that—after thousands of years of using horse carriages to move stuff around—in just fifty years, one person would be able to transport more cargo than two hundred horse carriages in one day, without a single horse. What an amusing, scatter-brained dreamer!

Since those days, all kinds of amazing innovations have improved labor productivity in agriculture and industry, way beyond anything once imagined. But now the world has more than seven billion people, and labor is no longer scarce. There are hands aplenty—fourteen billion of them, actually. All want work. Nature, however, is no longer infinitely huge. Resources are getting scarcer, and the prices of long-term commodities such as energy, minerals, and foods are rising.[9] Even more important, the natural sinks, particularly for carbon dioxide and the nitrogen cycle, have been overloaded.[10] The sky, oceans, and soils cannot take more pollution. But the good news is that green growth can solve sink overload and create jobs by eliminating the wastefulness through radical resource efficiency while employing more people. The next wave of innovation is the shift from brown to green growth, which builds on a shift from maximizing labor productivity to maximizing resource productivity.

Today, if someone claims that fifty years from now, we may get ten or one hundred times as much value created out of the same timber, oil barrel, or kilowatt-hour, they'd be ridiculed. "Not possible!"

Yet that's what leading businesses and investors are currently doing. Take lighting. Having a coal-fired power plant—or a diesel generator—make power, then transport it through the grid to a home where it lights an incandescent bulb, is an insanely wasteful way of meeting the need for indoor lighting. It's about 99.2 percent waste from fuel to light. Shifting to daylighting systems with light channels and better windows eliminates most of the daytime need. Then, at night, the home could be lighted with an LED bulb running on battery-stored power from a solar panel, improving the resource efficiency of the whole system by a factor of more than one hundred. And it's profitable for the end user, if not the incumbent utility.

When it comes to heating or cooling, green designers are creating passive houses rather than conventional houses, reducing energy use by nine-tenths. Then, by putting solar panels on the roof, the building could become net energy positive, generating more power than it consumes, since it now uses so little.

Personal transport is changing, too. Fossil-fuel cars are remarkably wasteful. Eighty-six percent of the energy in the fuel is wasted as heat in the engine and exhaust and never reaches the wheels. Less than 1 percent is actually used to move the driver and possibly a passenger. And then the typical American car spends 96 percent of its time parked, with only 2.5 percent of that time spent in the productive use of driving. The rest is used up in traffic or finding parking.[11] You've no doubt heard people say this system can't be improved much. But overall green growth approaches demonstrate that it's relatively straightforward to improve personal transportation by orders of magnitude over the coming decades. Tesla's Elon Musk has done more than any other to shake up the complacent car industry. Cities and towns are getting smarter about providing public transport options to replace cars. And electric bikes can be a hundred times more efficient than an SUV while also improving health and keeping the air cleaner.

On the food front, at least a third of the food produced around the globe is wasted one way or another. Reports show that nearly half the food produced in the United States and Europe never gets consumed. Food waste also leads to loss of natural resources since food production is accountable for 80 percent of deforestation, 70 percent of all the freshwater consumption, and 30 percent of greenhouse gas emission.[12] The amount of food lost and wasted every year is equal to more than half of the world's annual cereal crops (2.3 billion tons in 2009–10).[13] Upstream, rather than ruining soils when producing food, newer farming practices such as low-tillage can increase the amount of carbon in soils, making agriculture a net sink, not a source of carbon pollution.

There are positive stories in industry, too. The world's concrete industry is responsible for 5 to 8 percent of total world emissions. But with new processes and materials, like geopolymers, the process of manufacturing concrete may become near carbon-neutral with existing technologies.[14]

In the energy arena, solar and wind have seen costs fall over the last decades, while oil, gas, and coal have seen them grow. Renewables are outcompeting fossil fuels at more and more locations around the world. This makes for a fundamental shift in the energy markets, something the traditionally conservative Citibank calls a clear "energy Darwinism," where the unfit are heading for extinction. This is not through idealism or state intervention. It is through pure competitive force that huge, older coal plants are being decimated, and replaced by millions of nimble, low-cost solar panels and wind turbines.[15]

And don't forget about the Internet. If the Internet were a country, it would today be the sixth largest emitter of carbon pollution, mainly through its operation of computer centers and infrastructure. It is responsible for around 2 to 3 percent of world emissions.[16] But leading providers, such as Google and Apple, two of the world's three largest companies, have already committed to supplying *all* facilities with renewable energy and are providing the early signs of a fully renewably powered Internet.[17]

Taken together, there are incredible opportunities to improve over yesterday's wasteful practices. Green growth is the story about the profitable realization of these opportunities in the coming decades. It takes leadership, investment, and dedication to break the old mental patterns of fossil or brown growth. They seem "natural" and "reasonable," reinforced by recent tradition, cut-and-paste approaches, established culture and institutions. Costly subsidies for these outdated, wasteful practices must be curbed, and government regulations should support green growth.

Some are strongly against climate science and increased carbon taxes. But nobody's in favor of waste. Bringing about such a resource revolution is beyond doubt the greatest business opportunity of the twenty-first century. The main challenge is measuring, targeting, and implementing the resource efficiencies at all steps through the economic system, since the wastefulness of materials is invisible to those decision makers who rely only on monetary information. Gross profits can grow while ecological footprint diminishes.[18]

The one-liner is that *Green growth is smart, while brown growth is soooo twentieth century.* This is a narrative that will be attractive to businesspeople, entrepreneurs, technology optimists, politicians, and an economically minded audience. If climate resisters can be swayed, it will be through better solutions, not by more climate science reports. Many of the steps we could take to mitigate the problem, we need to take anyway. Is food waste noble? Is using energy smarter bogus? Can you find someone arguing that cutting energy costs is a bad thing? Someone who argues that wasting water is good, or that doubling resource efficiency is counterproductive, or that investing in ways to protect ourselves is foolish?[19] This is a much more attractive story to sell than the apocalyptic climate story.

## The Well-Being Narrative

Martin Luther King had a choice when standing in front of the crowd gathered before him at the Lincoln Memorial in 1963. As Futerra, a sustainability consultancy, pointed out in its *Sell the Sizzle* report, he easily could have leveled accusations and stirred the anger of his followers. The civil rights struggle had been long and hard, injustice was severe, and King and others had encountered death threats and danger. Striking back would feel just. Focusing on the bigotry, partiality, and discrimination would have been easy. The microphone was on. Silence was spreading through the crowd. Now what to say? Under immense pressure, he began with the words we all know today: "I have a dream . . ."—setting an inspiring ideal that lives on to this day.

"That's how it works. When you're faced with hell—you sell heaven," concludes Futerra.[20] You flesh out the story of where we need to go, in a manner that makes people really want and long for it.

Narratives like this focus on happiness rather than apocalypse, and depict the kind of society we want to live, laugh, and love in, and leave when the time comes. They are stories that emphasize well-being, social justice, and generosity as the new wealth.

Modern, industrial societies have been fabulous at generating wealth for people, and in particular for those lucky few who control most of it. Billions of humans have also been lifted out of destitution and hardship. Robert E. Lucas, Nobel Prize–winning economist, argues that the real impact of the industrial revolution was that "for the first time in history, the living standards of the masses of ordinary people have begun to undergo sustained growth . . . Nothing remotely like this economic behavior is mentioned by the classical economists, even as a theoretical possibility."[21] But maybe the worst criticism you could direct against modern society is that after achieving this basic level of material wealth for some, it does not further distribute or improve the well-being of its members.[22]

Since the 1970s in Western societies, most measurements of happiness and quality of life show no or very little improvement. Since then we've doubled average income per head, and doubled it again. Still no improvement in happiness or well-being.[23] All this coal-digging, forest-trashing, soil-wrecking, atmosphere-altering, ocean-acidifying frantic development, wealth accumulation, and competition . . . to what end?

Already in 1931, the eminent economist John Maynard Keynes spelled out this conundrum in the form of a hundred-year scenario to 2030, which is so well written it is worth quoting at some length:

> Let us, for the sake of argument, suppose that a hundred years hence we are all of us, on the average, eight times better off in the economic sense than we are to-day . . . Assuming no important wars . . . the economic problem may be solved, or be at least within sight of solution . . . Thus for the first time since his creation man will be faced with his real, his permanent problem—how to use his freedom from pressing economic cares, how to occupy the leisure, which science and compound interest will have won for him, to live wisely and agreeably and well.
>
> When the accumulation of wealth is no longer of high social importance, there will be great changes in the code of morals. We shall be able to rid ourselves of many of the pseudo-moral principles which have hag-ridden us for two hundred years, by which we have exalted some of the most distasteful of human qualities into the position of the highest virtues . . . The love of money as a possession—as distinguished from the love of money as a means to the enjoyments and realities of life—will be recognized for what it is, a somewhat disgusting morbidity, one of those semi-criminal, semi-pathological propensities which one hands over with a shudder to the specialists in mental disease. All kinds of social customs and economic practices . . . we shall then be free, at last, to discard.[24]

Now, eighty years later, developed countries are—as Keynes forecast—around eight times better off, but still struggling with his challenge: How to live wisely and well with our wealth? Our politics, particularly economic policy, is still stuck in the industrial age scarcity mind-set, where GDP growth is the holy grail of national development, as proved after the Great Recession of 2008.

The new story is one of improved life satisfaction and well-being, which now can be measured with psychological tools as scientifically as the GDP. This is a story that calls for the rearrangement of political priorities: human quality of life first, which is inevitably linked with our surroundings, the more-than-human world in which our lives play themselves out. The soul can't be happy while the biosphere is crumbling.

In short, the story envisions a future society in which people live with greater well-being, together with others, learning, giving, and caring. And as consumption measured in money is slowly growing, the ecological footprint will simultaneously be diminished, until it's inside the planetary boundaries.

I have a dream of an astonishing diverse plurality of cultures for the year 2050—all in which friendships, networks, organizations, learning, storytelling, and the arts flourish, in ways specific to their place and geography. City centers are highly walkable and bikable with millions of meeting places for chats and fun. The buildings are designed with care for the location, have passive ventilation with better air and uplifting day lighting. There is flirting, gossip, philosophical cafes, street theatre, peaceful protest marches, farmers markets, marathons, and rock concerts a plenty. The cars hum quietly around and do not spread toxic compounds from combustion engines, except for the occasional retro shows where noisy Formula 1–type events draw the petro-nostalgic with great beer and barbecue. Markets and trade are vibrant, and treated more as conversations of value than as efficient mechanisms of price equilibrium. Corruption, surveillance, and discrimination are being exposed by transparency initiatives and kept to a minimum. The food is short-traveled, healthy, highly varied, and incredibly tasty, and we will happily pay its full cost. The jobs are green and stimulate well-being through personal mastery and acknowledgment for work well done. A livable minimum wage that makes the freedom of social mobility more than a cliché. There is greenery everywhere in sight, on streets and buildings, so sparrows and hawks have recolonized inner-city rooftops.

When someone manages to describe a society that a strong majority of us long to live in, then things can start to happen, and happen fast. But if we have no idea of where we're heading, we certainly will end up somewhere else. Or as Lao Tzu said: "If you do not change direction, you may end up where you are heading."

This happiness narrative gives a sense of direction where the doom narrative only says no. It removes the need to believe in climate change or not. This is the direction we want to go in anyway. One famous cartoon by Joel Pett in *USA Today* captured the gist of this sentiment by depicting a climate skeptic exclaiming: "What if the whole climate issue was a hoax, and we ended up creating a better world for nothing?"

What is your story of where we ought to go? What is a society worth living—and dying—for? If we don't know, we surely won't get there.

## The Stewardship Story:
## The Greening of Religion and Ethics

A historian of medieval history, Lynn White at Princeton, wrote a now classic article on the relation between religion and environment back in 1967. He came out with some stark comments on the effects of Christianity on humankind's relation to nature:

> Especially in its Western form, Christianity is the most anthropo-centric religion the world has seen . . . Man shares, in great measure, God's transcendence of nature. Christianity, in absolute contrast to ancient paganism and Asia's religions . . . not only established a dualism of man and nature but also insisted that it is God's will that man exploit nature for his proper ends.
>
> In Antiquity every tree, every spring, every stream, every hill had its own genius loci, its guardian spirit. These spirits were accessible to men, but were very unlike men; centaurs, fauns, and mermaids show their ambivalence. Before one cut a tree, mined a mountain, or dammed a brook, it was important to placate the spirit in charge of that particular situation, and to keep it placated. By destroying pagan animism, Christianity made it possible to exploit nature in a mood of indifference to the feelings of natural objects.[25]

This was immediately contested by a series of Christian thinkers, which in turn ignited a rethinking of Christianity's relation to the natural world. New fields of study like environmental history, environmental ethics, and eco-theology were created and stimulated. Out of this movement has grown a new and powerful story, the ethical stewardship story, which goes along these lines:

> Humans are created by God and placed on the earth, which also is equally created by God. This makes the earth holy, and humanity's proper role is to be caretaking stewards of it.

More and more Christians are finding quotes and principles in the Bible replacing the traditional human *domination* over creation from Genesis with the equally biblically based language of human *stewardship* of creation. God the Creator has entrusted His creation to the stewardship of humanity. It's His will that we treat it gently and respectfully.

Several of the world's religions are turning the same way: According to Islamic scholar Seyyed H. Nasr of George Washington University, "Islam makes no distinction between the natural and the supernatural" and holds that the Qur'an was revealed to the entire living world, not just to humans. And Asian religions offer additional subtle notions of balance and harmony through Taoist, Confucian, Jain, and Buddhist philosophies. Scholars of the world's major religions are reexamining their scriptures, rituals, and doctrines to bring forward those elements that support an ecological awareness and a changed sense of humanity's place within creation.[26] Most Native American peoples have always held such a view of the natural world. The words *respect* and *thanksgiving* sum up the Native American attitude, according to Oren Lyons, Faithkeeper of the Haudenosaunee (Six Nations Iroquois Confederacy).[27]

Jewish investor Yosef Abramowitz sees the failure to transition to a more sustainable economy as an act against faith in shared human responsibility. As he more eloquently states, "Because of our corruption and our greed we are jeopardizing the miracle of creation every day."[28] Therefore, installing solar panels and LED bulbs to replace diesel generators and incandescents becomes something akin to religious duty.

After a summit on sustainability at the Vatican, Pope Francis made the religious case for acting to limit climate change, stating that "Creation is not a property, which we can rule over at will; or, even less, is the property of only a few: Creation is a gift, it is a wonderful gift that God has given us, so that we care for it and we use it for the benefit of all, always with great respect and gratitude." Francis also said that humanity's destruction of the planet is a sinful act, likening it to self-idolatry. "But when we exploit Creation we destroy the sign of God's love for us, in destroying Creation we are saying to God: 'I don't like it! This is not good!'"[29]

There is evidence that such views are spreading among American Catholics and evangelicals. A University of Maryland poll of religious Americans found that 76 percent support a treaty to counter global warming. If such an agreement is ever reached, religious Americans say they would stand by it with conviction. Of the fifteen hundred people surveyed, 57 percent say that violating such a treaty would be morally wrong. About 17 percent see it as a sin, requiring atonement to avoid everlasting consequences. The same 76 percent of respondents believe preventing climate change is an important goal. Among them, 32 percent say it falls within their obligation to protect God's creation. A bigger group, at 44 percent, don't think of it as an obligation but feel that it's important to defend against rising temperatures nonetheless.[30]

For those who do not associate with any particular religion, there is a parallel narrative of moral and ethical obligations. You could call climate change the perfect moral storm, since it brings together three major ethical challenges. First, it is a truly global phenomenon. Once emitted, greenhouse gas emissions can have climate effects anywhere on the planet, regardless of their source. This yields skewed results—the poorest and most vulnerable countries and people are those that have emitted the least historically. The second challenge is that current emissions have profoundly intergenerational effects. Those who are still not yet born have no voice in our conversations. The third challenge is that our ethical tools are underdeveloped in many of the relevant areas, such as the moral value of nonhuman nature and the extent to which we have obligations to protect penguins and polar bears, salamanders and salmon, and similarly with unique places, or even nature as a whole.[31]

The conservative worldview also fits well with this story: It is vital to conserve nature. Being a responsible steward of nature is key to keeping traditions alive and tending the landscape so that the next generation inherits the land in as good or better condition than it was in when we inherited it. There is a proud tradition of conservation and respect for the natural environment within conservatism. Trusteeship and the acknowledgment of a shared responsibility to conserve and protect the natural environment are therefore all embedded deeply at the heart of traditional conservatism.[32]

One way to move ahead with the environmental ethics is putting it into international law. Polly Higgins, a London solicitor-turned-activist, has been arguing that ecocide—the equivalent to homicide but applied to critical ecosystems—should be seen as an international crime. She's inspiring a global movement that is strongly proposing an amendment to the international Rome Statute: "Ecocide is the extensive damage to, destruction of or loss of ecosystems of a given territory, whether by human agency or by other causes, to such an extent that peaceful enjoyment by the inhabitants of that territory has been or will be severely diminished."[33]

This ecocide narrative says that people, governments, or corporations that irresponsibly kill off species or forests, or endanger climate stability, are committing crimes against humanity. There are countless examples of ecocide happening around the world from the Athabasca Tar Sands to extensive logging of the Amazon. Such cases could then be tried in front of an international jury. If the jury decided so, these acts would be illegal. Individuals, not just corporations, can be held personally liable. They should be tried as soon as possible, and not be left for judgment in the afterlife.

## The Re-Wilding Story: Bring Back the Wildness

Re-wilding is a story of nature's amazing capacity to rebound. It tells of the resilience of birch and pine, the dogged persistence of badger and fox, of dandelion and vine, the dark and voiceless worlds of algae and fungi. Many of the interdependent mammals, birds, and corals may be vulnerable, living precariously close to the extinction cliff, but nature is also wild and robust, and swings back if given the smallest crack in the concrete. Witness the dandelions. And of course, nature is not something out there. We're fully inside it, from breath to bones. We live inside the chain from crops to compost.

Activists have been clear about what people should not do. We have been urged to consume less, travel less, live not mindlessly but mindfully, eat broccoli but don't tread on the grass. Without offering new freedoms for which to exchange the old ones, activists are often seen as puritans, spoilsports, and complainers. Climate messengers have been very clear what we're against; now we must tell the story about what we are for: a love for the wonders of the wild. Rejoicing for the wind and all things winged. Sharing stories about the beauty of places where we live, the art and culture there, and the amazing four-legged and rooted beings we share this land with. In the book *Feral*, George Monbiot proposes an environmentalism that "without damaging the lives of others or the fabric of the biosphere, offers to expand rather than constrain the scope of people's lives." He writes,

"It offers new freedoms in exchange for those we have sought to restrict. It foresees large areas of self-willed land and sea, repopulated by the beasts now missing from these places, in which we may freely roam."[34]

Just outside Oslo, a wolf mating pair was recently, in 2013, observed moving into the city forest. Their choice for a new home was quite deliberate; the delicious roe deer that occasionally feed on gardens and fields here, in addition to the buds, scrub, and herbs of the forest, had become among the most abundant in the country. But the heavily populated Oslo area hasn't had any breeding wolves for more than 140 years. I'd never heard or seen a wolf in the wild before, so when I was invited out by one of Norway's leading nature photographers, Ulf Myrvold, who's making a film about the wolf's return, my heart started drumming.

After following some fresh tracks in the wet snow, we were amazed at how close to human settlement these animals lived—without anyone inside their living rooms knowing that there are wild hunts going on just beyond the field. The forest that we walked into was heavily managed, mainly spruce and pines. But the multitude of tracks in the wet snow turned the forest floor into a living storybook. We saw where the roe deer had strolled, where the wolf had run across them, and where the pack had come in from the field.

We decided to sleep under an abandoned shed, which would hopefully retain our all-too-human smells, but still had a good view to the border between fields and forest. We settled in, listening for wolves but hearing nothing. Hours of shifting clouds. Finally something moved under the clouded dark night. No; too small, "just" a fox. More long nothings. We startled at a howling sound but it was a tawny owl making its "ho-ho-hoo" in the distance. The moon fought its way through the cloud cover and started brightening up the snow, spreading the uniquely silvery light of Arctic winter nights.

Then, half an hour before midnight, a howl shattered the silence. Every cell startled. Hairs stood out stiffly inside my soft down sleeping bag. The howl was deep-throated, gutsy, unflinching. And very, very close. We looked at each other incredulously. The previously very domesticated farmer's field, the regimented forest, the worn-down, tired shed—all this was now electrified. The howl came again. And yet again, ending on an even deeper tone, cracking up from tones into low-frequency rumbling. It was easy to determine the direction of the sound. Just over there. But why couldn't we see him or her? The wolf kept itself covered behind the trees at the edge of the field. Probably checking out the tracks in the area, circling around our old shed, maybe moving inaudibly on the night-frozen snow just behind us.

Utter silence settled back in. But between the pines and the stream in the center of the field, between the snow and hillside behind us, everywhere the wildness was now thick in the air. Palpable. The planted spruce and the cultivated field, at daytime so domesticated and calm, had been transformed into a breathing, alien ultra-presence, each snow crystal bristling with excitement under the fresh moon rays. Our ordinary rural landscape had been transformed, while only half a mile to the north people were brushing their teeth after switching off their favorite reality show. Sleep came, but just before dawn I was watching again. Something was moving on the other side of the field. Wolf? Fox? Badger? Impossible to say as long as it was half hidden behind some leafless trees along the stream. After breakfast, we inspected the field. No doubt; there were wolf tracks in the snow. My perception of the "timid" Oslo city forest changed forever. Wildness abounds, just under the surface of the commonplace.

On the other side of the world, smack in the middle of Los Angeles, feasting on deer and roaming the chaparral-covered slopes, a mountain lion has started prowling Griffith Park for the first time in more than a century.[35] Bringing back the lions, foxes, and bobcats is necessary for reenchanting the city, says James Gibson, a sociology professor at UCLA.[36]

What does this have to do with climate? Biologists clearly tell us that ecosystems with higher biodiversity and old-growth forests have better resilience and carbon-capture capacity than monocultures.[37] Re-wilding our cities and agricultural landscapes—and interconnecting these biodiversity areas with wildlife corridors—increases the likelihood that the ecosystems in which we live can both adapt and mitigate the coming climate changes. Top species in the more-than-human food chain, such as wolves and whales, elephants and tigers, seem to create trophic cascades that ripple "down" through the ecosystem, increasing and maintaining the dynamic dance among species around some kind of fluctuating optimum. We need that wildness in order to remain fully human in the coming centuries. As the essence of that self-willed wildness, the presence of wolves is exactly what we need more of.

Stories of re-wilding counter the laming apocalypse tale by reinforcing that we're not going to reverse our damage, but nature can, if we'll only reduce our interference with it and let it do its work. It's a team really: the wilderness and us. And I'm all for teaming up with nature, as humbly and intelligently as we humans are capable of.

## The Inevitability of Stories

What story creates meaning for you? Which will you tell? Of course there are more options than the four main narratives I've outlined above.

How we tell the story is just as important as the archetypical story pattern we use when talking about the facts. Storytelling is a discipline, as much as science is. The hope lies in integrating the two. Rather than hoping that just numbers and facts will speak for themselves, we must each integrate science with storytelling, and not leave the job to stressed-out, catastrophe-prone journalists. In this work there must be room for humor, emotion, visualization, point of view, climax, surprise, plot, drama. Above all, make it personal and personified.[38] There is plenty of psychological evidence that information that is vivid, salient, and personified can have a larger impact on people's behavior than information that is statistical and abstract.[39]

There is no reason to expect that experts in physical and climate sciences should be experts in storytelling. Actually they are trained to take all crumbs and remains of the personal out of the equation. They are peer-reviewed into killing any subjective statements or touchy-feely elements from their articles. For them to communicate effectively the important findings of their objective research to the broad public, however, the personal and a broader imagination have to come back into the process.

I fully respect scientists who say that they don't want to meddle in the story-telling business. Their calling is exploring and explaining earthly processes. But their thorough research lends them unique credibility as messengers within their area of expertise, much more so than any clever, professional storyteller. Therefore it is my hope that more than a few will want to step forward as communicators to the public. This will be in addition to their all-in work as scientists within marine biology, atmospheric chemistry, meteorology, statistical methods, satellite measurement, glaciology, oceanography, soil sciences, geology, watershed science, forestry, and a whole host of similar sciences. They don't need to become master storytellers. That's hard. A more modest starting point is mainly to become

## Stories: Tell Better Climate Stories

Avoid apocalypse narratives, and instead tell stories about:
- Green growth.
- Happiness and the good life.
- Stewardship and ethics.
- Re-wilding and ecological restoration.

When telling stories, make them:
- Personal and concrete.
- Vivid and extraordinary.
- Visual; "show, don't tell."
- Humorous and witty, with strong plot and drama.

aware of the underlying storyline in whatever is being said. Did we just trip into that apocalypse narrative again?

Better storytelling can overcome our deepest barriers, particularly the barriers of denial and identity. Climate messages have been unpalatable because they—in their apocalypse form—evoke fear, guilt, and helplessness. The dominance of the disaster narrative in more than 80 percent of news[40] bears witness to a disastrous failure of imagination among communicators, including journalists and editors.

Any story that tells me that my identity and lifestyle are wrong and destructive will be subconsciously resisted. We all reinterpret the facts in light of a favorite story that sustains our understanding of ourselves as valuable citizens, whether liberal or conservative, religious or atheist. For those of us who find ourselves stuck in the moral conundrum of the climate doom story, passive denial offers an easy way out.

Only through new, attractive stories that we want to identify with will we start to reconsider the scientific facts. Different audiences need to hear different stories. The apocalypse story only reinforces the old barriers, while stories of economic revitalization (green growth), happiness, social justice and the good life (vision), stewardship (aligning ourselves with higher values), and re-wilding (teaming up with the self-healing forces in nature) all offer us a way around the most deeply seated barriers in the minds and hearts of modern citizens. Luckily, the four are in no way mutually exclusive: We need green growth for better well-being, becoming stewards of the land we help re-wild.

One question that has been driving me in this book is whether people are—as some claim—locked into short-termism. What would it take to move beyond that trait in terms of climate—if this is even possible? The new emerging strategies seem to converge toward an answer: We will willingly shoulder the extra burden of acting long-term if we have a community with such social norms, supportive frames in which to make decisions, simple nudges for everyday actions, and/or some kind of grand stories about where we want to go that give us a sense of common purpose. It's not that we are incapable of acting for the long term; it's just that the conducive conditions haven't been there.

This leads, finally, to the need for better metrics—the new signals to steer by.

# THIRTEEN

# New Signals of Progress

How do we imagine progress? When there are imaginative stories of where we want our societies to go, there arises a need to know whether we are moving toward that future. Having meaningful signals and feedback is key to keeping up the momentum of transformation.

On the national level, the gross domestic product (GDP) has for more than sixty years been the yardstick by which we measure and understand economic and social progress. In the beginning this signal served its purpose. However, it has failed—according to economist Joseph Stiglitz and colleagues—to capture what makes a difference in people's lives and contributes to their happiness, such as leisure, security, income distribution, and a thriving natural world.[1]

The economic "hero" of GDP accounting would be a terminally ill cancer patient going through expensive treatment and a costly divorce. This kind of life really gets the dollars circulating, but it may not be exactly what we're looking for. Pumping all the oil and gas to burn it as quickly as possible, killing all the forests for timber, and hauling up all the fish in the sea would also initially register as an increase in the GDP. Thus, the indicators that measured human societies' progress from the first half of the twentieth century have proven insufficient for this century. Even the Human Development Index (HDI) fails to reflect the state of natural resources or ecological conditions, says economist Partha Dasgupta.[2]

Climate communication messages usually give numbers such as tons of $CO_2$ emissions, inches of sea level rise, and average surface temperatures in Fahrenheit. The problem is that when seen along with economic growth figures such as GDP, housing prices, and stock exchange indexes, this approach gives people no way of knowing whether climate concerns are in the process of being addressed or not. Are our climate policies and responses good enough, a little weak, or hopelessly inadequate? The climate problem is a so-called wicked or diabolic problem; it is complex and interrelated, and there is no good way to know when it is solved—or if it ever will be in our time scales.[3] This sets us up to fail. Frustration follows.

In order to maintain engagement in societal transformation, there has to be a way to speak of and get feedback on any type of turning in the right direction. Without such feedback, there is little learning and less motivation—only increasing confusion and helplessness. The challenge is to translate our societal response into signals in an easily understandable way. Can we really measure if we're changing in the right direction (the derivative of the curve), whether we are doing too little, just enough, or too much?

This need to know whether we are making headway could be made more personal and relevant if results could show not just what is happening on a global and national level, but also whether each person, company, city, and state is doing at least its fair share or too little. We need indicators that visualize complex information in compact ways that are clarifying and useful to public decision makers as well as to our social networks. Then we could discern how we relate, quarterly or yearly, to the larger direction of societal change.

The choice of signals has a huge impact. Management guru Peter Drucker famously said: "What gets measured, gets managed." Let that sink in for a minute. Robert Kaplan, the founder of the Balanced Scorecard, flipped it over: "If you can't measure it, you can't manage it!" And strategy adviser Igor Ansoff sighed: "Corporate managers start off trying to manage what they want, and finish up wanting what they can measure." The last quote captures the current situation: We have leaders and media *that want what is measured*, and that is first and foremost the economic growth signals such as jobs, consumer confidence, and GDP (the more the merrier).

For new signals to be used they must be perceived as meaningful and not just mere numbers. This can only happen when signals are closely knit to the larger story of societal transformation, such as the stories of green growth, well-being, stewardship, or re-wilding. Good signals make it possible for people to make sense of current policies in relation to what they care deeply about.

It's critical that climate communication is designed in close combination with indicators for measuring societal progress, so that feedback motivating meaningful action is given. For instance, at the individual level people are more inclined to recycle than better insulate their homes. But of the two, the more effective for emission reduction is energy conservation, so an indicator visualizing progress on energy conservation would therefore make more sense for climate than measures of recycling.[4]

All of this is why the field of research and development of indicators and signals is bustling with trial and error right now.[5] Many climate communicators are unaware of this field, and of the benefits of letting these signals accompany the conventional global climate figures (such as $GtCO_2e$/year, $CO_2$ ppm, or average surface temperature). However, in order to connect to the supportive framings and stories, the indicators should measure the success of solutions, not focus on global problems.

Among the promising new indicators are a handful that can track our societal responses to the climate challenge: a green growth indicator called GEVA, two well-being signals called Happy Planet Index and Integrated Wealth Indicator, an ethical stewardship indicator called the Kantian Climate Policy Indictor, and finally a re-wilding indicator called the Nature Index.

## Signals of Green Growth

Green growth indicators measure the change in economic activity from one year to the next: Is our growth greener than the previous year in both quantitative and qualitative terms? They assess how much closer we are to a green economy, a future end state where the entire economy operates fully within the planetary boundaries, and no longer pushes ecosystems or the air beyond their limits.

A good green growth signal will therefore show the change over time of the relation between any economic activity and its impacts on the more-than-human world. The best starting point is the relation between climate gas emissions and the value added of a company or other economic entity. Value added is sales revenue minus external purchases, quite similar to gross profit. Put simply: How much emission do you generate for each million dollars of added value? Let's say a corporation emits one hundred tons of $CO_2$ from its internal operations for each million dollars of value added. How many opportunities for emission reductions must this company realize while generating more value per year to have sufficient smart green growth? What is green *enough*?

Well, there is a widespread consensus that the whole world must at least halve emissions by 2050 to have a fair chance of avoiding too much runaway climate disruption. This could allow us to meet the famous 2 degrees Celsius (3.6 degrees Fahrenheit) goal from Copenhagen 2009. From 2015 to 2050 is thirty-five years, meaning that on average emissions must drop at least 2 percent each year. But all companies, of course, want to grow their profitability. Some grow fast, others slower. Historically companies grow their operations on average 3 percent per year measured in value added. If we factor in that growth, then each company must reduce its emissions per dollar of value added by (2 + 3 =) 5 percent per year, each year from 2015 to 2050. Then the company would be doing its fair share to halve climate emissions.[6]

The idea is: more goods (value added) with less bads (emissions). Impossible? No, the good news is that consultants from McKinsey & Company have calculated that cutting 2 to 3 percent $CO_2$ per year is actually profitable for most US corporations at least up to 2020. And the Alliance to Save Energy has set the goal of doubling energy efficiency by 2030, saving three hundred billion dollars yearly and creating 1.3 million jobs doing it. The cheapest fuel is the one you don't use. "Negawatts" are often cheaper than megawatts. Reduction of fossil fuel costs goes straight onto the bottom line.[7]

Thus, to stay green (and not just be window posing or greenwashing), our hypothetical company must by next year get from 100 to 95 tons of $CO_2$ per million dollars of value added. The year after that it can still grow its revenues and profits while achieving another 5 percent cut in the green growth indicator—down to 90.25 tons per million dollars. And so on, for series of years, to 2050. If we add up all value added from all economic entities, it equals the total GDP of a nation. Thus, if all corporations (including government bodies and foundations) did the same, then the entire nation would be part of green growth, too.

Each corporation—regardless of industry, including oil companies—can thus become part of the necessary and profitable transition to a green economy. We can measure this transition through GEVA, the Greenhouse Emissions per Value Added signal (see figure 13.1).

While the GEVA measures the green growth of corporations, we can also measure the green growth of a city or a whole nation. A green nation is one that achieves around 2 to 3 percent growth in monetary GDP terms

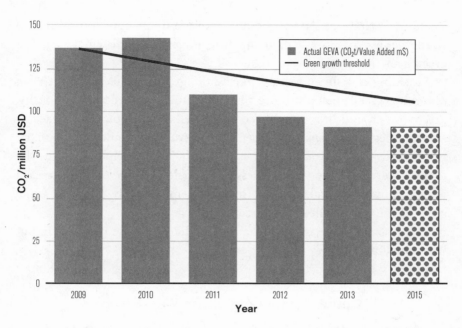

**Figure 13.1.** The Greenhouse Emissions per Value Added, or GEVA, indicator shows whether an economic entity—here the company Tomra—is achieving real green growth or merely greenwashing. This figure shows that they are successful in reducing their own emissions relative to value added and are ahead of their 2015 target. They are a truly green company. Source: Tomra annual report.

while at the same time reducing emissions more than 2 to 3 percent per year. In other words, it grows its productive system in the direction of a stable green economy within planetary boundaries. Sweden and Switzerland are two countries that have achieved green growth in the 2000–10 period.[8] As more people realize that this is fully possible, fun, profitable, and precisely measurable, we can accurately pinpoint who's performing and who's not—by integrating economic and ecological performance into one signal that is applicable on corporate, city, state, and national levels.

Consequently, with an understanding of what is green *enough*, future climate communicators could say: "Atmospheric $CO_2$ concentrations have further increased and just passed the unprecedented 410 ppm level. This is worrying. However, the two last years we're on track to achieving a green economy by 2050. So if we keep this pace, we'll turn around in time. But Arizona is slipping while Texas is leading, so Arizona better shape up!" This shifts from a doom frame into an opportunity frame, strengthening the sense of competing for nationwide efficiency and green growth.

Tackling climate emissions may be the greatest challenge right now, but going green includes more than just emissions reductions. Ideally, green growth indicators would account for all of our footprint impacts— from species extinctions to nitrogen-cycle disruptions (in which too much runoff leads to acidic or dead zones in water). The principle would be the same: An ecological footprint assessment would measure progress toward reducing our ecological footprint (including emissions) by 2 percent while growing the value-added by 3 percent per year.

That covers some example signals on the economic green shift, but what about the turn toward the good life?

## Counting Happiness and Well-Being

Taking all things together, how happy would you say you are these days . . . ? 1—very happy, 2—happy, 3—not very happy, or 4—not at all happy?

This is a pretty common item in research and measures of happiness. Some people might think that this method is too simplistic, but the answers to such simple questions appear to be surprisingly meaningful, particularly if you can combine them with more detailed and varied questions and measures of freedoms and capabilities. A whole science of happiness, or positive psychology, has sprung forth in recent decades. We now know in which countries people are the happiest, and can explain why some people are happier than others. It is now fully possible to objectively measure the development of subjective well-being in different segments of the population.

Climate communicators ought to be aware that the time is ripe for our measurement system to shift emphasis from only measuring economic activity to also measuring people's well-being. Changing emphasis does not mean dismissing GDP. But emphasizing well-being is vital because there appears to be in rich countries an increasing gap between the GDP data and what counts for common people's well-being. Above a certain level of income per person, there seems to be no more gains in happiness for further growth in average income. This means working to complement measures of market activity (GDP or revenue) with value-added measures centered on people's well-being and that capture sustainability.[9]

What we want are happy people on an ecologically healthy land: a high quality of life *and* low emissions; happy people and a bustling multitude of four-leggeds, feathered folks, creeping-crawly things, and all those under-wordly wonders bustling in the humus and soils. A happy life doesn't really have to cost the earth.

One attempt to capture the combination of human happiness and ecological impact is the Happy Planet Index (HPI). It tracks national well-being against resource use and was introduced by the UK-based New Economics Foundation in 2006 to measure the ecological efficiency of supporting well-being. The index is weighted to give progressively higher scores to nations with lower ecological footprints.

One criticism of subjective happiness self-reports is that valuing well-being goes beyond people's own perceptions to include measures of their functioning and actual freedoms. In effect, what really matters are the capabilities of people, that is, the extent of their opportunity set and of their freedom to choose the life they value from this set. This spans education, health, political freedom, and environmental health. Therefore some have argued for a more integrated view of progress.[10]

In order to measure a broader sense of sustainability, they say, we need indicators that inform us about the change in the factors that matter for future well-being. We need to grow several "stocks" at the same time: natural, human, social, and produced capital. Natural capital is the quality of the ecosystem. Human capital is the knowledge, competence, and health of the population. Social capital refers to trust in relations and stability of institutions, while produced or manufactured capital is all buildings, machines, roads, grids, and so on.

Luckily, great progress has been made in recent years, particularly by pioneers such as Partha Dasgupta, in measuring these four capitals and seeing them combined. He has been central to developing the Integrated Wealth Indicator (IWI).[11] An example is shown in figure 13.2 for the 1990–2008 period.

What the IWI shows is that most countries have in this period used and drawn down their natural capital (mostly the negative bars, below the line in figure 13.2). But those with a net-positive IWI have converted the natural to more human and produced capital. The IWI thus shows the

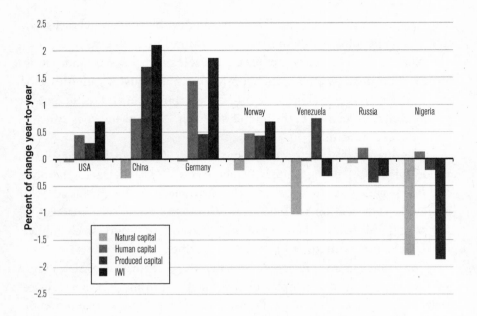

**Figure 13.2.** Integrated Wealth Indicator for 1990–2008. Annual growth rates (per capita) disaggregated by capital form and sorted by IWI.

composition of the capitals and how they are balanced. But some countries (those to the right) consume more natural capital than they generate in other capitals. This is not a sustainable path to well-being, since their natural capital is crumbling, without corresponding gains in human and produced capitals.

## Does an Ethical Indicator Make Sense?

Ethics may be hard to measure, particularly on a national level. Some philosophers would also strongly object to measuring "ethical progress."

But today, as of this writing, there is still no international emissions treaty binding for all countries. It is highly uncertain if there ever will be one. Still, each country could do something. Indeed, from an ethical point of view, each country *should* do something, even if there is no binding treaty forcing action. Immanuel Kant pointed out that we ought to do our

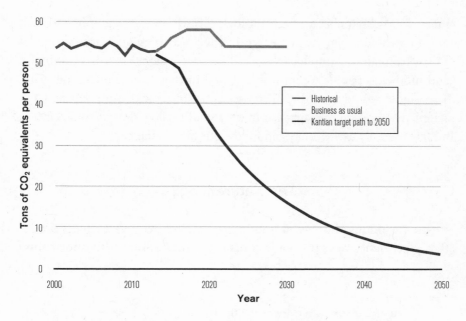

**Figure 13.3.** Kantian Climate Policy Indicator for Norway: Is the country reducing climate pollution *as if* a fair, effective global treaty were already in place? This means following the target line to 2050.

duty not because it serves our interests or furthers our desires, but for its own sake.[12]

Following Kantian ethics, each nation ought to act *as if* a sufficient international treaty were already in place. This would require each nation and state to carry out all greenhouse gas mitigation projects to the point where emissions per person are equal to the global average footprint in a world with sustainable emissions by 2050. A sustainable level may be on the order of one or two tons of $CO_2$ equivalents per person. Today, the worldwide average is seven tons of carbon $CO_2$ equivalents per person per year.[13] In the richer countries, it is between ten and fifteen. This measure is relatively easy to track, and can thus serve as an indicator for what is good enough at the national level.[14] If some countries began mitigation work, it would increase the probability of more joining in. Finally a treaty binding for all could come into being. (See figure 13.3.)

This Kantian approach could work even better at the individual level. Each person would get the same right to emit a certain amount but no more. The sustainable and ethical long-term limit is the aforementioned one to two tons per head per year. Thus, an individual accounting of $CO_2$—from all our purchases—could become a signal that is meaningful and motivates people to become engaged in their own climate footprint. This signal would work best if it were integrated with shop labels, bank statements, and mobile apps, so that we could follow our personal carbon budget as easily as we today can follow our bank balance.[15]

## The Nature Index

If you are among those who find the re-wilding story particularly attractive, you might prefer to couple climate communication with information

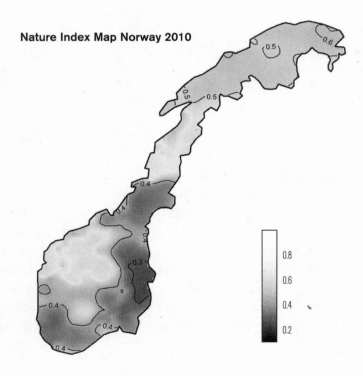

**Figure 13.4.** The Nature Index. The map depicts Norwegian open lowland biodiversity for 2010, where dark areas represent low biodiversity relative to its original state, and light are high and intact biodiversity. Source: The Norwegian Environment Agency.

### Signals: Integrate Climate Communications with New Indicators of Progress

To support new stories, we need new indicators to see and give feedback on progress:
- Greenhouse emissions per value added (green growth story).
- Happiness, well-being, and integrated wealth (vision story).
- Kantian ethics planetary boundaries (ethical story).
- Ecosystem health and nature index (re-wilding story).

on how the various species are doing. What is the status of nature's biodiversity? Ecosystems with high biodiversity seem more resilient, and also increase well-being for people who connect with the land there.

The *Ecosystem Millennium Assessment Report* from 2005 described a grim outlook for many of earth's ecosystems. A global IPCC-like body for ecosystems has been established.[16]

The Nature Index (figure 13.4) is a way for the Norwegian government to monitor the overall quality of different ecosystems by using color on the map to represent values from 0.0 to 1.0. Here, zero is severely degraded biodiversity (dark), while one is intact, full-level biodiversity (light). Biologists who know and study the ecosystems give each area a score for bird, mammal, fish, plants, and other natural assets. The different biological assessments are then integrated into a number representing the ecosystem's overall health, and expressing the richness and biodiversity of all its plants and animals. Just as the value of houses is determined by real estate experts, the Nature Index score for each acre is determined by ecosystem experts.

Looking at Nature Index maps or animations over time, we can easily get a visual impression of how ecosystem health is developing for those areas that we live in, or are interested in. For instance, Yellowstone National Park's biodiversity increased following the reintroduction of wolves.

## Our Response to Signals

How we respond to signals, or indicators, depends on how accessible, interactive, and relevant they are. "Just numbers" don't mean much. But if we can make the signals vivid and interactive and available through social media and social norms, we may see them come alive among the public. When connected to stories, they create meaning. Getting the signals of our progress right is absolutely essential for the long-term success of climate communications. Otherwise the global climate data will have no impact on social decisions. Climate data such as atmospheric concentrations are global, aggregate figures only and create disempowerment and disillusionment, since no person, group, or nation is able to influence them. To see measurements of what we as citizens can influence is very motivational. It brings the issue into our sphere of influence. And this influence must make a difference in people's lives, to the local ecology, and to the transformation of society in response to climate change.

# —PART III—

# Being

## INSIDE THE LIVING AIR

*Can you hear the sound of life*
*in the roaring of the creek*
*in the blowing of the wind*

*That is all I want to say*
*that is all*

—*NILS ASLAK VALKEAPÄÄ,*
TREKWAYS OF THE WIND

—FOURTEEN—

# The Air's Way of Being

*We are lived by powers we pretend to understand.*
—W. H. AUDEN, *"IN MEMORY OF ERNST TOLLER"*

Too focused on our daily busyness—the thinking and doing parts of our lives—we've grown oblivious to the air, how we air-condition it, how we breathe it, and also how the air breathes us.

As philosopher and author David Abram says, "What is climate change if not a consequence of failing to respect or even to notice the elemental medium in which we're immersed? Is not global warming, or rather, global weirding or even global burning, an unintended consequence of taking the air for granted?"[1] Tackling this spreading crisis is about technical, economical, political, and strategic considerations for sure. But more fundamentally, we also need to deepen our view into being and beauty, ethics and aesthetics. With the frantic tilts and sways of global weirding at hand, we will be forced and humbled into rethinking humanity's relationship with more-than-human nature and with the wild air itself.[2] In the climate crisis, the air is making itself felt, troubling us, and thus increasingly coming to our attention again.

Only occasionally, when the foul local smells of untreated pipe emissions from a large plant, the stench of city sewage, dust storms, or the smog of a great city attacks our eyes and noses do we suffer the air's condition, and turn—not to the air itself, but to local technical solutions. Economists turn to the Kuznets curve, a "law" that says that as countries get richer, they eventually start to reduce pollution. They will gradually be able to afford filters that remove the particles. Or they'll move polluting industries farther away, or build higher smokestacks that spread the pollutants farther. The idea is: Use nature to get rich first; deal with the emissions later.[3]

But the gaseous emissions that cause global warming are not like the smoke and other particulate matter that becomes what we call air pollution. Greenhouse gases travel much farther than local pollution, with deeper and longer-lasting impacts. They spread stealthily and invisibly to become part of air itself. They linger and subtly change our very existence now and for centuries to come.

It may be worth taking a step back and asking: What *is* the air? What is this gaseous bubble that envelops us, this lively system that we are part of?

When speaking of the atmosphere in the climate change discourse, we tend to think about it as something out there, or up there. It is a mix of molecules. In this mix there are nitrogen, oxygen, a little argon, and then the now infamous carbon dioxide and other greenhouse gases. Of which there are now too many so-called ppms, they say.

This is the primary, scientific framing around our climate discourse: The air is a mix of gases, each with certain physical and chemical characteristics. This way of understanding the air arose with the chemical revolution in the Enlightenment period of 1789 to 1830.[4] But to our breathing bodies, the air is much more than just a mix of molecules passively floating around, with energy input from the sun, interacting with ocean and cold space. We have long viewed the air only through the chemical, reductionist view; now it's time to consider other ways of seeing, feeling, and relating to the air.

## Earth's Skin

First, we can imagine the air as the earth's skin. It is amazingly thin, compared with the size of the planet and the cosmos it shields us from. Just a fine, flimsy film. The breathable air is only about five to seven miles thick, a fragile wrapping around a massive green, blue, and white ball, seventy-nine hundred miles in diameter, spinning in near emptiness. On a clear night we look right through it. Far thinner than the skin of an apple compared with its diameter. Underneath it lies ocean and rock, and upon the rock, lies a wee bit of soil and greenery. Yet inside this unsettled, fluctuating film, between a rock and a cold place, all of life is protected and nourished. It stops destructive cosmic rays that would otherwise destroy our skin and our fragile proteins and genes. It insulates and regulates temperatures in a range that is suitable for water and life. Mediating between blue ocean and black eternity, it carries all the billions of tons of water in the clouds needed to replenish the soils. It fills the rivers, stirs the waters, drives waves and ocean currents.

How can we speak differently about this air? It's not just "the climate" or "the atmosphere," an object *out there*. Imagine we move in slowly from outer space. You could then envision the air as a gradual thickening of Nothingness into Being. It is a mercurial borderland where absolute emptiness transitions into a huge but thin gas bubble hosting a mingling of cosmic plasma with liquid water and solid earth.

But where does the air stop? It's not just out there, far up above us. It is also inside us. It impacts us here and now. It literally fills my lungs, and yours, too, from the first breath to the last. Its elements travel in our blood, before returning through the alveoli to the outside. It fills all the pores in our skin, making us semipermeable. Yet the air isn't just what we breathe into our lungs, briefly visiting before we exhale it. It is also our primary link to the world. It fully envelops us, from the soles of our feet to each hair on the top of our head, from the day we draw our first breath to beyond our death. It holds us gently, with a benign embrace without which our bodies would fall apart. To be born is to enter the air. To be is to be in the air.

This generous and stable medium—this invisible context to all our lives—has, as an integral part of the earthly biosphere, maintained itself in a relative stable balance over the last ten thousand years, while we have walked around on our two legs, invented agriculture and writing, and expanded our fledgling civilizations. Before that, the air had sustained favorable conditions for life for about 3.8 billion years.[5] It is into this thin layer that all of modern humanity now is collectively pissing by dumping gigatons of off-gases from ancient carbon through our various combustion devices. We have altered the composition of the air without thinking much, so far, about the outcomes. This vital rethinking and acknowledgment have only hesitatingly begun in the last couple of decades.

The air that you are now breathing in through your nostrils, the air through which your eyes gaze at the words composing this sentence, has been dramatically altered to a composition it hasn't seen for three million years or more.[6] Climate change isn't really distant. It is as close as that: your nostrils. We do know that the air—for us two-leggeds—holds potentially very violent powers. With increasing frequency, as its well-calibrated heat flows get tilted, the atmosphere unleashes torrential rains, freakish super-storms, enduring droughts. These can and will push large forests, watershed ecologies, and plains—as well as all the birds, so many trees, butterflies, mammals, and amphibians—outside their known historic boundaries. Some creatures will move, adapt, shift, and mutate, while others will disappear forever. And with them, many of us may see our lives uprooted, flooded, dried out, or impoverished.

The empirical and conceptual schools of psychology applied in parts 1 and 2 can only take us so far in understanding how we think and act in response to the facts of climate science. They don't address much about *who we are and how we exist in the air.* They tell more about thinking and doing than about being. And they tend to overlook the fact that we are not observing the atmosphere somehow from the outside but are ourselves intimately permeated by it. Whether we are conscious or unconscious about the air's influence, we are always already immersed within it.

## Awakening the Sleeping Giants?

To some, this may appear less scientific, less rigorous, less hard-minded, more philosophical and more far-fetched. Some more analytic or tough-minded readers may choose to stop reading here. To me, though, shifting how we think and act in relation to climate change is one thing; exploring how to more fully "be" in the world around us is another. This exploration isn't mystical or soft-minded, but rather a highly necessary extension of the psychology of climate into more existential and phenomenological domains.

In order to make sense and meaning out of our being here, humans have always turned to stories. Grand stories of gods and ancestors and heroes. Inadvertently, climate change also touches upon such archetypal stories and figures. In the old days, ancient Norse peoples perceived angry gods inside weather events. There was Thor with his hammer—the god of thunder who still has a day of the week (Thurs-day) named in his honor. Greeks knew when Zeus was out with his lightning bolts. The winds were persons, too. Zephyros, the west wind, was the gentlest of the winds—a lush wind seen as the messenger of spring.

Gods, however, don't enter into the vocabulary and discourse of modern scientists. What were long ago discussed as acts of gods are now called hurricanes and arctic jet streams. Meteorologists speak of vast and expanding Hadley cells of global wind patterns that are probably drying out the subtropics; or they talk of the thermohaline circulation—also called the great ocean conveyor belt—which hugely impacts the earth's climate. They do not speak directly of the Old Norse Midgard Serpent that grew so large, he was able to surround the earth and bite his own tail. But that's what the great ocean conveyor belt still does.

If we listen with a mythological ear, there are deep stories underneath the modern climate messages: Humans have altered the composition of the ocean (acidification) and air (now more than four hundred ppm $CO_2$) so that the large weather systems are changed and upset from the relative stability of previous millennia. Humans have interfered with the larger arctic wind currents and contributed to melting glaciers, which are even shifting weight on some continents, maybe even creating more earthquakes, and sending atmospheric jet winds outside long-term bounds. The polar vortex, a normally peaceful high stratospheric pattern, has started shifting. These sleeping giants,[7] from both below and above, seem now to be awakening and starting to act wildly and uncontrollably.

Reviewing the climate science of tipping points is outside of the scope of this book, but some results indicate that when the Arctic multiyear ice is soon fully gone, we *may* see a meltdown of methane hydrates in permafrost and seabeds. This *could* unleash enormous burps of methane from below, tipping the climate for . . . well, in human time scales pretty much forever. If the term *gods* refers to more-than-human earthly powers, then, yes, the god of ocean and rising sea levels, *Neptune*, and the god of the dark underground world of geology, marshes, and methane hydrates, *Hades*, and the gods of sky, cloud, and thunder, *Zeus* and others, seem to be getting increasingly pissed off. Such giants may be not only burping (and we're talking about seriously gigantic acid reflux burps here) but throwing major tantrums.

From a narrative point of view, there is not much difference between modern stories of ocean conveyors, methane burps, and polar vortices, and these ancient stories of Neptune, Hades, and Zeus—the gods of sea, underground, and sky. They are still there. Even if our language has evolved. We're still confronted by more-than-human powers whose ways of working we do not really fathom and are even less able to predict. "We are lived by powers we pretend to understand," as the poet Auden observed.[8]

It is one thing to hear a climate change speaker list all these factual changes in earth ecosystems, or read about these changes in the news, where each event is just another statistic, or digest the simple weather facts from NOAA's annual reviews. But it is something else altogether to let that information sink into us—sink past all our biases and barriers, past the socially conditioned attitudes, beyond the different cognitive framings. Then finally, maybe, the *meaning* of the facts starts to congeal in the soul. The larger story becomes more visceral and substantial inside. And with it, probably, grow lumps of worry, disbelief, and astonishment about the magnitude and speed of current climate disruptions.

How do we relate to all this as emotional and ethical beings? How do we live with the knowledge that within the lifetime of a little human baby born today, our common home, our *oikos*, is likely to be hotter than at any time for three *million* years? A world in which the long slumbering giants are now awakening?

# FIFTEEN

# Stand Up for Your Depression!

*"Depression" has made us aware of our culture's addiction to a manic superficiality in growth.*

—JAMES HILLMAN,
RE-VISIONING PSYCHOLOGY, P. 25

In addition to the thinking and doing, there is also something else. There is a deep sadness, a quiet undercurrent of grief that does not abate. It is difficult not to feel depressed by this onrushing calamity that will—with high probability—change most of the landscapes you learned to love while wandering as a child, the woodland paths where you walk your dog, and the butterflies, frogs, wild salmon, coral reefs, bats, and bees you might see along the way.

People, too, are losing their houses and livelihoods due to increasing floods, tidal surges, and heat waves. Moose, mosses, and monarchs are silently disappearing with nothing but a whimper, since their habitats or migration routes are deteriorating.[1] Some highly mobile species of birds, insects, and fish may be able to relocate. Others are trapped on mountain ranges or by human infrastructure such as roads, dams, or agriculture. More than half of all animals gone in the last forty years, says WWF. If there is nothing but indifference or even optimism in the face of these ongoing massive meltdowns of earth's wealth, the amazing beauty we've inherited, then "you may want to seek a psychotherapist," suggests ecologist Stephan Harding.[2] Rather than not feeling much about it all—the common explanation for apathy—we maybe feel too strongly, notes psychoanalyst Rene Lertzman. Since we don't mourn sufficiently, we're stuck with a repressed, cut-off despair inside.[3]

This more-than-personal sadness is what I call the Great Grief, a feeling rising in us as if from the earth itself at this time. This is an ancient idea in our culture, going back at least to the Roman poet Virgil, who wrote of "the tears of things," *lacrimae rerum*. It refers to tears inside nature itself.[4] Perhaps bears and dolphins, rivers and the acidifying, plastic-laden oceans bear grief inside them, too, just as we do.

The first time this Great Grief came to visit me, many years ago, I was lying on my back in the wooden house I was renting, relaxing on a rug, not far from the woodstove, which was silently munching away on some dry, crackling spruce logs in late October. The ceiling in this two-hundred-year-old wooden building was low and tilted. It was really a good time in my life: safe income, stable relationships, nothing in particular to worry about apart from drying my wet clothes in the warmth of the stove. I'd just been outdoors in the drizzling forest, walking with my Belgian shepherd dog. Then a deep, deep grief settled in me. Unannounced. Uninvited. As if arising from the soil itself, or descending from the heavy clouds above.

All I could think about were the species leaving us, the demolished forests, the impoverished humus, the washed-out soils, the disappearing corals, the First Nations and dispossessed driven off their homelands, the poached tigers in rapidly receding forests. The grief was not triggered only by climate change itself, but by the extended loss of beauty, diversity, and habitat. Would entire landscapes be devoured by opportunistic beetles taking advantage of the heat and the drought? Would we lose the myriad ways of coexisting in the air, where insects mingle with bird flight? And what would vanishing wetlands mean for the salamanders and frogs that enjoy the way water intertwines with the mud?

So many human languages—unique ways of singing and vibrating in the air—are going extinct as well. And many, many more-than-human languages—from the songs of birds to the ultrasonic beeps of bats to the huffs of bears—are dwindling away so rapidly in this planetary moment, inexorably drifting into oblivion and leaving not much more than frail human memories, stuffed skins in museums and some digital photos. Or maybe a DNA trail deep-frozen and stored in a lab somewhere.

Yes, even in this moment of intense intuition about the major madness in modernity's shadow, I am fully aware that humans have made progress. We've no doubt multiplied over and over. In industrialized countries, we now have great houses, lights, warm showers, touchscreen keyboards, and the latest life-saving antibiotics and anesthetics. All good stuff. A lucky few have gotten very, very rich. But the destruction and desecration! The sorrow of it all. Tears began to run down my cheeks, as if from the rain itself.

This Great Grief hits many of us, and it is not very articulate when it does. It expresses itself in a manner more emotional than cerebral. It is more like a heavy fog drifting through the soul. The gray epiphany does eventually subside, though, as most heavy rains do. Luckily, we get back on our feet. The forces that bring us there are more than we can personally change, so it would be unreasonable to assume we could fend off future envelopments. There will be times when we will fall into this sadness. What we *can* control is how we choose to get back up again. And how we find the grace to live with it.

## The Mordor of Eco-Anxieties

As signs of global warming accumulate, therapists report they're seeing more and more patients with eco-anxiety symptoms. Pamela Larsen, forty-one, a mother of two young girls in Oregon, gets a stomachache every time she looks up at the long-dormant volcano nearby: The glaciers at its peak have definitely been receding over the years. As the mountainside gets browner and browner from beetle-infested dying pines—evidence of climate change—the knot in Larsen's gut tightens.[5]

For indigenous peoples, land, plants, and animals are considered sacred relatives, far beyond a concept of property. If any are destroyed, the loss becomes a source of grief. Members of the Karuk Tribe of California have been intimately dependent on salmon and other foods from the Klamath River since time immemorial. When one Karuk saw the deteriorating water flows from drought and pollution, he said:

> I think particularly for indigenous people social, cultural, and community wellness reflects the ecological quality of their environment. So when the river's degraded, and it's liquid poison in some ways, and you're supposed to draw all your sustenance and your identity as a river Indian or a river person, then of course that weighs on you. It's like, you want to be a proud person and if you draw your identity from the river and the river is degraded, that reflects on you.[6]

Australian aborigines, Navajos, and any number of other indigenous peoples have also reported a sense of mournful disorientation after being displaced from their land. If you force them from their homeland, they feel the loss of heart's ease as a kind of vertigo, a disintegration of the basis of their whole life and their deep spatial knowledge of places. As climate disruption degrades their ecologies, the land disintegrating in front of their eyes, this can be experienced as yet another forced removal.[7]

The loss of heart's sense of ease with the land is not limited to displaced indigenous populations. Some years back, Murdoch University professor Glenn Albrecht began receiving desperate calls from residents of the Upper Hunter Valley, a six-thousand-square-mile region in southeastern Australia. They felt they were losing the land they lived on due to coal mining, dust, drought, and polluted rivers. He coined a term for this new sense of distress—*solastalgia*, describing the pain or sickness (*-algia*) caused by the loss of solace from one's homeland. Solastalgia exists when there is recognition that the place where one resides and that one loves is under assault.[8]

To make matters worse, the pollution impacts of large fossil-fuel-burning plants tend to hit the most vulnerable communities—black, brown, elderly, and those of low income. In the United States, coal's health risks affect African Americans more severely than whites, largely because a larger portion of African Americans live close to coal plants and coal waste dumps. Asthma attacks triggered by exposure to above-average pollution levels send African Americans to the emergency room three times as often as whites.[9]

Across different populations, psychological researchers have documented a long list of mental health consequences of climate change: trauma, shock, stress, anxiety, depression, complicated grief, strains on social relationships, substance abuse, sense of hopelessness, fatalism, resignation, loss of autonomy and sense of control, as well as loss of personal and occupational identity.[10]

Sufferers of such eco-anxieties may go through sudden, uncontrollable bouts of sobbing. And maybe worse than the sobbing, we feel dejected and doubtful of our ability to change anything. *What can little me do?* The overwhelming images of disaster can freeze us like rabbits in the headlights. Like weak witnesses to some cruel calamity beyond our capacity to even comprehend what's occurring. Rather than take it in, numbing ourselves so we don't feel too much of it is a common reaction. The psychiatrist Robert Jay Lifton has described this as a psychic numbing, when the level of suffering becomes overwhelming—as in large-scale atrocities or with the threat of nuclear war.[11] Gradually, our sensitivity and the heart's capacity to respond are worn thinner until nothing really awakens an emphatic response. We live as if anesthetized or drugged, our hearts numbed.[12]

The American writer Wendell Berry puts it this way:

> *It is the destruction of the world*
> *in our own lives that drives us*
> *half insane, and more than half.*
> *To destroy that which we were given*
> *in trust: how will we bear it? . . .*
> *To have lost, wantonly,*
> *the ancient forests, the vast grasslands*
> *is our madness, the presence*
> *in our very bodies of our grief.*[13]

· That our individual grief and emotional loss can actually be a reaction to decline of ecology is an idea that rarely appears in conversation or the popular media. Neither does it circulate widely in the discourse of contemporary society. It comes up frequently as anxiety about what world our sons or daughters will have to cope with. Many bring it privately to a therapist. It is as if this topic is not supposed to be publicly discussed.

Most prefer not to come to, or at least not dwell on, this Mordor of eco-anxiety, despair, and depression. One of denial's essential life-enhancing functions—the periodic benefits of which should never be underestimated—is to keep us more comfortable by blotting out this wintery darkness. We could call it the Ecology of Grief,[14] or—in the words of activist Joanna Macy—Environmental Despair.[15] But even if we may need some denial to keep us up and running, we may also need periodically to bring some light to this depressive domain. Otherwise we could end up trying to shut it out completely. Or it could finally get the upper hand and overwhelm with the darkness, leaving us blown off balance, burned out, and lost.

If, instead of avoiding this hurt and grief and despair, or aggressively blaming "them" (the industrialists, politicians, agrobusinesses, loggers, or bureaucrats) for it, we could lean into the feelings, we might discover something inside them. It seems, somehow, important to persist and get in touch with the despair that arises from the degradation of the natural world.[16] That's because contact with the pain of the world can also open the heart to reach out to all things still living. It breaks open the numbing. Maybe there is community to be found among like-hearted people, among those who also can admit they've been inside the Great Grief feeling the earth's sorrow, each in their own way. Maybe there are some poems to be made, petitions to be signed, torches to be carried, demonstrations to join with dance and music, or some other social good to come from this encounter of the raw heart with an abused part of the earth. Maybe even a renewed philosophy of life: What does it want from me? How does this give meaning to my life?

The highly original cultural psychologist James Hillman stated that he believes in the transformative power of depression: "Through depression we enter depths and in depths find soul. Depression is essential to the tragic sense of life . . . It brings refuge, limitation, focus, gravity, weight and humble powerlessness. It reminds of death."[17]

So maybe we need more, not less, deeply felt despair in our culture at this time—particularly among those in power, like financiers and decision makers. But are too many running from it, seeking an escape by flipping from depression to manic action? Is not the entire titanic modern energy and growth system a symptom of ignoring the limitations and humility that humans need to reacquire?

Depression is a state of mind that brings exactly such a sense of limits and humility. Through it we may discover that we're not on top of the world. Rather than being masters in control, *depression teaches that we are fully dependent*. We're vulnerable, depending on well-functioning air, oceans, and fertile soils to nourish our own bodies and our feverishly growing but fragile civilizations. Maybe depression can slow down the rush so we can re-search for less destructive ways of coexisting with the earth's ecosystems.

Hillman speculates further: "The depression we're all trying to avoid with being manically busy could be a prolonged chronic reaction to what we've been doing to the world . . . Perhaps the way to begin the revolution is to *stand up for your depression*."[18] A way to ease the manic-depressive cycles we experience is sometimes to actually enter the depression more fully, and imaginatively—by speaking poetically in dark phrases, ranting a rap of our anger, making art from it, building a house on Dark Mountain. Then this newly shaped depression can moderate our subsequent manic phases somewhat: By bringing some expressions of the depression into the manic phase, and vice versa, the extreme swings may moderate each other.

On the cultural level, standing up for your climate depression would mean giving voice to the feeling of limits, despair, gravity, slowness, rather than manically going shopping, increasing energy capacity and consumption, or getting GDP growth rates up again as quickly as possible. Ours is a heroic can-do culture, where there is a more or less permanent, hubristic denial of the depressed side of the soul. The implicit message is: *You've gotta have solutions. Don't come here unless you have a solution.* And to be depressed and weak is a disgrace, sick, and distasteful—hence we hear all too often, "You should get some medication for that."

## Humbled from the Pinnacle—Shaping the Dark

One fish said to another: "This thing called water; do you believe in it?" In the West we still swim in the water of a prevailing and pervasive background philosophy. But we are mostly as oblivious to it as we assume fish are to water: That philosophy arises from the belief that we are the dominant species, the species that wields the power and is entitled to exploit other species, which are on earth solely for human benefit. This arrogance is a specific mode of thinking and feeling. Many today call it anthropocentrism. This assumption that humans are the pinnacle of earthly life elevates us to a superior mode, in which presumed rational control and technology seem to give us mastery over the earth.

This mind-set has led to the geological period some now call the anthropocene,[19] a new geological epoch in which human influence is the strongest force dominating the face of the earth.[20] But climate disruptions expose the folly of human self-declared superiority. This hubristic we-are-on-top view is effectively punctured by eco-depression. In an unguarded moment, darkness sneaks up from the soul's depths undetected, flip-flopping the grandiose self-confidence over to the abysmal. When inside our consciousness, this depressive inner voice seems to be whispering—with an ominous sense of humor: "The question is not whether there is intelligent life on Mars, but whether there is intelligent life on earth? No convincing evidence of true intelligence has so far been found in the life-form dominating today . . . But then this life-form will be gone soon anyway. In another million years or so the earth will have recovered from this evolutionary error called humanity."

I've seen this line of reasoning in many places. From the dark writings of Schopenhauer and Nietzsche to the Norwegian pessimistic philosopher Peter Zapffe, to the writers, artists, and thinkers today involved in the Dark Mountain Project.[21] It crops up in the dystopian sentiments of movies such as *The Matrix* and in Internet forums and discussions on climate change impacts.

The challenge is to be able to bring such feelings from the innermost private rooms to the public floor, without slipping into the narrative of a literal doomsday coming soon (the "end is nigh" story) or becoming anti-human or misanthropic. How, though, do you do this? You need to let the despair out, unashamedly but also nonaccusingly. Listen to how it expresses itself in both your heart and mind, by being fully immersed in it when it appears. Avoid the temptation of shifting into a sermonizing, moralizing, pontificating, or demonizing mode, and simply stay faithful to the gray images of desolation and destruction, understanding how they weigh heavily on your shoulders, heart, and feet. Let the emotion of "I give up" flow through you, but don't identify with it.

Thus, the humbleness may—at the least—be heard. Maybe some others will recognize and identify with it, and will experience release when listening to you. We feel better when sharing the genuine sorrow of our soul confronted with the losses that climate disruptions are bringing, through sharing our worry for animal species we love, our sense of despair when floods strike the already marginalized, when drought-stricken farmers give up, when Peruvian and Nepalese villages are wiped out by melting glacier mudslide avalanches or Bangladesh farmers flooded with nowhere to go. In all of these responses we can remain true to our being as compassionate earthlings; we can witness the suffering of earth's human and more-than-human beings without pointing the finger of blame at someone.

This is not to say that there are no blame-worthy initiatives and blame-worthy people fronting them. Or that we should love all and accept all. Digging out tar sands, mining away mountaintops for coal, or creating vast plantations in the heart of the rain forests are just a handful of the ecologically insane practices ongoing today. And there are plenty of vocal, powerful people who ruthlessly defend and expand them despite volumes of documentation on their destructiveness. We may need to stand up against such practices hard and long to quench them. My point here is that there must *also* be room and space where the genuine despair may be expressed and heard. Maybe my anger needs to cry without being impatiently and prematurely pushed and bullied into positive thinking, quick fixes, and social movements. Yes, we must make haste. *And* yes, we must make haste slowly, *festina lente*, with the kind of deep questioning that allows the heart and the soul time to follow.

From a psychological point of view, it helps to let this sorrow be shaped, sung, wept over, and maybe even honored and celebrated with a ritual, a deep song, or something, before you enter into the more action-oriented modes. You can then enter the fray with a cleaner mind, so to speak. You're not throwing your unprocessed sorrow, depression, and anger into the face of a coal mine employee, a lumberjack, a local fisherman, an energy executive, a pilot, a captain of industry, a stockbroker, a petroleum analyst, an international climate negotiator, or simply ignorant public servants, lumping all into a guilt-laden "them" that the "righteous us" are fighting against. This splitting dynamic of us-good versus them-evil is widespread in Western culture after two thousand years of Christianity and monotheisms. The kind of binary thinking that only recognizes opposites like heaven and hell, right or wrong, left or right, has taken its toll. It furthers the disease of polarization and fundamentalism. The results show up in the culture wars and political partisanship that is pulling the United States apart.

The good-versus-bad dualism has unfortunately given much activism a moralizing shadow. It has created unnecessary fronts, tensions, and backlashes. The climate message then comes across with so much urgency and anger, so hard and laden with unprocessed despair and guilt, that this emotional background is heard more loudly than the urgency or content of the message itself. Greenpeace, for instance, enlists a desperate Santa Claus to vent its anger at misbehaving politicians who don't want to save Santa's home.[22] And Irish environmental writer John Gibbons wrenches his fervent heart that:

> transnational corporatism, turbocharged by neoliberal economic theory, has reached its global apotheosis and, freed at last from the constraints of either human moral agency or state oversight, is in the process of destroying the very civilization it purports to

serve. Climate change may not feel like your fight today, but that too will change. Make no mistake; this is, quite literally, the fight of our lives.[23]

Anger can be most appropriate when reacting to outrageous conditions. Feeling a mix of grief and anger about the state of the climate, rivers, and forests is only appropriate. Writing or speaking from our anger can bring fire into otherwise dull, blunt, and numb expressions. There is intense anger, for instance, in Picasso's grayish, toned-down painting *Guernica*. And funneling anger into action can make for enduring, effective motivation. Yet when anger flows through us, it is challenging to handle with some level of grace. It's tempting to spit it out in the face of others. But giving anger free rein in a splurge of acting out, so that it spills over into accusation and assault, often backfires. Do accusations and calls for an all-out fight such as the one in the quote above, right or wrong, help win people's hearts—or make them turn away from the message and messenger? Let's say that *they* are wrong—and *you* are right! Scream it to them. Does that make you feel better? Sure. But does it help climate or policy much? The problem is not that what you say is wrong; it is polarizing on an issue where we need to move beyond.

Even when messengers in touch with their despair and anger try hard to keep anger in check and stick to the facts, many audiences have become hypersensitive to accusations of being somehow "bad." Due to an already large but unconscious deposit of inner despair and guilt about overconsumption and climate emissions, listeners will easily feel that the climate messenger is pontifical and condescending. They will then be looking for ways to get back at the one who is communicating her despair about the future climate.

The feelings of despair and anger are not due to an absence of goodness or decency in our lives, or to a lack of natural and cultural richness and creativity around us. Rather the depression is a response to fact that all the good things we value are under threat—and that "the attempts made to protect them have been weak in comparison with the power of the destructive forces now being mounted against them," writes psychoanalyst John Steiner.[24]

This is the situation that the early psychoanalyst Melanie Klein considered to be at the heart of our experience of depression as toddlers: that we become closely attached to such good qualities and other beings, and "*at the same time* are unable to protect them from destructive attacks."[25] But conventional psychoanalysis has mostly focused on the human infant–mother relationship. Can these dynamics be extended from the human mother to Mother Earth? To the human infant, the mother is early in life experienced as part of him- or herself. Can we regard Mother Earth the same way, and connect ourselves to all her cows, dogs, clouds, or patches of land? Why not?

## Entering the Borderland

Whenever I bike past a certain roadside plot of land, I get a knot in my stomach. Biking by not long ago, I noticed that the dried-out, half-dead pine seedlings and grasses there had just gotten a lifesaving late-summer rain shower. But still, they were growing so slowly. Precariously. Their roots could not go deep, since the soil that had once coated the bedrock had been excavated a few years back. On, oh horror, my instructions, to make a horse-riding paddock. When the paddock plans fell through, the site was supposed to be restored with new soil, but the excavating company didn't comply with my repeated requests. Now nature tries to recover by itself on the patchy remains of gravel here and there.

I like to envision myself as a do-no-harm activist, but I've participated in the ripping up of Mother Earth. Depressing. There is an open wound there under the skies on about a hundred square yards of land. It was made with the best intentions for teaching eco-psychology and equine therapy and then restoring the area. But both intentions failed, and what's left is just this wound with the half-dead saplings where there used to be good soil and lush green growth. How to deal with the sorrow of such attempts that derail?

I traveled to Santa Fe, New Mexico, to discuss core climate psychology questions like these with eco-philosopher David Abram and also meet psychotherapist Jerome Bernstein. Bernstein is an elderly gentleman, educated as a conventional Jungian analyst, but he also works with elders of the First Nations of New Mexico. Over recent years he has discovered that more and more, patients are bringing with them into the psychotherapy room not just their relations to mothers, fathers, and spouses, but also their relations to the wider natural world. In his book *Living in the Borderland*, he relates a series of clinical cases where the patient's feelings for and dialogues with trees, birds, and cows seems as vital and real to them as human relations.

One patient in particular, a woman called Hannah in the book, helped convince Bernstein that our consciousness is now actually extending into the more-than-human world, into the soil, oceans, and the air. He describes his decisive "conversion" to this view:

One day, a year or so into the work, she arrived at my office very distressed. The feeling was that the cows were being taken to slaughter. I pursued the standard approach of suggesting that she was projecting onto the cows, i.e. how she saw her life circumstance in the plight of these cows. She went along with me for a time. But then she protested in frustration: "But it's the cows!" I pointed out to her that her response was identification with animals she experienced as abused. She acknowledged the truth of my interpretations. She began to talk about all the animals in the world that exist only as domesticated beings, and their sadness. And again she burst out: "But it's the cows!" After that last protest—by now at the end of the session—I became aware in myself of Hannah's distress and her identification with the plight of these cows. And I also became aware of a different feeling in the room. The feeling was attached to Hannah, yet it was separate from her. It seemed of a different dimension. It was a new experience for me.

Some weeks later, Hannah recounted how she had gone for a long walk in the country and was followed by some stray dogs. As she described the experience, the room filled with pain and remorse. I asked her what she was feeling. Again we had a go-round like the one with the cows. And again she acknowledged her projection onto the dogs. But this time, out of character for her, she became angry—so angry that she took her shoe in her hand and hit the floor with it. "You just don't get it!" she shouted, and slammed the floor again with her shoe. "It's the dogs!" It was as if she were saying the dogs were projecting something onto her. The urgency of her tone and her uncharacteristic anger jolted me into the realization that my standard interpretations were not enough and somehow off the mark. Something other was happening in the room . . .

As I listened to Hannah struggle to articulate her emotions, I did "get it." It was indeed the cows. I realized that what Hannah was telling me was precisely the same message the native elders and healers were teaching me—and what my own unconscious was telling me through my dreams: Everything animate and inanimate has within it a spirit dimension and communicates in that dimension to those who can listen.[26]

After this session, Bernstein started to call this area the Borderland— not borderline, as in the psychiatric diagnosis. It consists of the psychic space where the hyperdeveloped and overly rational Western ego is in the process of reconnecting with its split-off roots in nature. The Borderland is the land where our consciousness merges with the landscapes around us. It is that in-between realm where the world speaks to humans, and humans to the world. It's a wild realm that we sometimes fall into and sometimes cannot find. Sometimes it's sparkling magic and at other times brings deep sorrow and despair over what is happening to the forests, the glaciers, or the creatures, as was happening to Hannah.

But why would—or should—we welcome or even visit the eco-despair? Experience in psychotherapy teaches us that we may find a type of nourishment there that we couldn't find anywhere else. Just as many plants need autumn and winter in order to flourish the following spring, we humans may need to connect to this darker ground of our being before we again can flourish into full action. There are seasonal cycles in the soul, just as there are in temperate valleys and plains. There can be a lot of psychic energy bound up in our despair and in our active avoidance of it. If you feel the despair pressing on you at times, it may be because it wants to be let in, to come more fully into your life. Though it may feel terrifying and repulsive to let yourself drop into the black feeling that "it is all for naught," the experience can be both cathartic and transformative. The feeling is not just a personal one; it flows from the world and through you. *It* comes to visit you. If you open the door. And it can even connect you back with the earth.

Aldo Leopold, one of the founders of ecological ethics in the United States, addressed the phenomenon this way:

One of the penalties of an ecological education is that one lives alone in a world of wounds. Much of the damage inflicted on land is quite invisible to lay [people]. An ecologist must either harden his shell and make believe that the consequences of science are none of his business, or he must be the doctor who sees the marks of death in a community that believes itself well and does not want to be told otherwise.[27]

California writer Derrick Jensen writes about the despair in a more angry voice:

> The most common words I hear spoken by any environmentalists anywhere are, "We're fucked." Most of these environmentalists are fighting desperately, using whatever tools they have—or rather whatever legal tools they have, which means whatever tools those in power grant them the right to use, which means whatever tools will be ultimately ineffective—to try to protect some piece of ground, to try to stop the manufacture or release of poisons, to try to stop civilized humans from tormenting some group of plants or animals. Sometimes they're reduced to trying to protect just one tree . . . But no matter what environmentalists do, our best efforts are insufficient. We're losing badly, on every front. Those in power are hell-bent on destroying the planet, and most people don't care.[28]

Paul Kingsnorth of the Dark Mountain movement posits that a "cult of optimism" keeps us from wanting to hear bad news:

> Every time you tell people about climate change or any other horrible thing that's happened already or is coming along, people just don't want to hear it. We've got this whole global movement of climate change denial now which is an incredible thing really, psychologically. Millions of people out there, busily working away pretending it's not even happening . . . you can't give out any kind of honest green message on a wide scale in a society in which people are as addicted to material prosperity as we are here. It's just not possible. And that leaves Friends of the Earth and Greenpeace and the Green Party and all the rest of them in a very difficult position, an impossible one really.[29]

Having been deeply in despair, as described by Bernstein, Leopold, Jensen, or Kingsnorth, can also have value if it kills off the ego's naïve hopes. Despair sets us free by letting us acknowledge the futility of holding on to our wishful thinking and heroic illusions. We are not in control, and things feel like going from bad to worse. We must let go, surrender into this blackness, into grief. But we also need to move through it.

## On the Far Side of Grief

The psychiatrist Elisabeth Kübler-Ross worked closely with people in grief and found that many pass through a series of five stages in how they relate to loss, death, or other deeply troubling news.[30] Though the model was developed to assist dying and grieving patients, it nicely describes the challenge of facing climate disruption, too.[31]

The first stage of grief Kübler-Ross identified was *denial*. Many simply refuse to accept the message of cancer: "I feel fine." "This can't be happening, not to me." This is the same kind of self-protective defense we mount against the climate message. In the second stage denial often becomes mixed with *anger*: "It's not fair!" and "Who is to blame?" Actually, we're pissed at those blind people in power who are hell-bent on destroying the planet, just as Derrick Jensen described above.

The third stage involves *bargaining*. We try to modify, mollify, sweeten, gloss over, or insure against the coming death, divorce, or calamity. Bargaining, in terms of climate change, may mean that we simply find ways to adapt; "A little warming can't be that bad!" or "Well, I'll buy an extra air-conditioner before next summer." "I'll move higher up on the hill." "I'll drill a deeper well." "Let's move farther north."

Fourth comes *depression and despair. I'm so miserable, why bother with anything? Whatever I do has no effect*. It is usually futile to attempt to cheer up someone in this stage. It is an important time for grieving that must be processed. We sometimes alternate among denial, anger, bargaining, and despair.

Finally, we arrive at the fifth stage, *acceptance*: People come gradually to terms with their own or a loved one's mortality, sickness, or loss. Importantly, in acceptance we can also find resilience. The bland bright optimism recedes, a darker, tougher version of hope is rebuilt, and new existential strategies are developed.

Depression is an all-too-familiar state to me and my fellow climate change activists. Every weather abnormality comes with a sense of dread.

If it is the case that the truth will set you free, then the hard truths about the climate disruptions may set you up for taking "anti-depressants for the rest of your life. It's at this point that we lose people" in general, writes Daphne Wysham.[32] Denial suddenly starts looking more attractive than acceptance.

But there is something stirring in the Western psyche, too. Something new. Another, more direct mode of contact. Just as we seem to be losing all these amazing creatures, we cannot help but reach out to them. There are more and more Borderlanders through whom the land is speaking. There are more and more souls that cannot help but be deeply touched by more-than-human beings. The disappearing species and glaciers and rivers and pines seem to reach out to us, too. Despite the many who do not care, full of resistance, expanded awareness grows in many others.[33]

The paradox is that behind all that despair, as we travel through it toward acceptance, we may find a renewed way of caring for the land, air, ourselves, and others. We may shift gradually from helpless depression to a heartfelt appreciation and re-engagement. Sometimes our hearts are numbed because of stress, overstretch, habituation to destruction or cultural prejudice. When despair hits, when tears flow, those numbed hearts suddenly crack open. Something deeper, wider, wilder peeks through and into our closed-off consciousness. Through the air that breathes in us we're suddenly reconnected to the world. And, as if we have feathers, wings, or sails, somehow the air doesn't just nourish us, but also lets us float and be lifted by its powers.

## Surrendering to Our Wings and Roots

When I think about how we come back from despair, I recall one late-August day when I'd been struggling to get a new clean-tech start-up company on its feet. I was also deeply distressed about my relationship to my beloved. I packed my backpack with a heavy heart and stressed-out mind. I headed up the mountain behind my cabin. It was heavy uphill going, which felt very good, and as I was working my way up a steep cliff an eagle hung in the air above me, pretty close, just above the precipice. For some reason, I decided that right then and there I had to pull off my backpack and get out my diary—as if there were something the eagle wanted to say. I wrote an ode to the feathered raptor, inspired by its keen sense of navigation, its ability to effortlessly circle the dark mountain and draw the uplift it needs from the rising air around it. It was riding the thermals to get where it needed to go; it wasn't stressing itself with hard wing strokes.

The sitting, the writing, the watching gave way to a sense of flow, of surrendering to larger flows. As the eagle slowly disappeared behind the misty far skies, it left me with a heart that still resonated with those wings, somehow uplifted, too.

There is a strong, strong capacity in human beings not just for despair and anger, but for love as well. Love for eagles and pines, willows and birches, clear streams, deer and foxes, cranes and eagles. Also the love for one another, for the love of what's local—from cultures, friends, fellows, and houses to pubs and beer. We even have a love for foreign cultures and mountains and glaciers that we may not have visited. Through story and images, traveling in the imagination of novels and films, we may connect intensely to other beings far off. And we may actually find nourishment and uplift there, as if we have wings. It's not our own power, ego power, but powers from the land itself, born on the wind, that let us connect to the land through the forces of imagination.[34]

There may be no bonding as strong as the attachment we form to the land while walking on children's feet. The landscapes of your childhood are forever contained in the body's memory. I remember very well the tree where I used to climb, a large birch with a dual stem and many lovely branches on which to step, sit, or hang. That old tree stood in Ålesund, my hometown on the northwestern coast of Norway some forty years ago. Still, today, I can recall the differences between the softer, yielding blackish branches and the thicker, whiter branches. Some branches I could step safely on. But others would bend or break. There were a couple of large stones just below the tree, and once, I fell from a thin branch and broke my right hand. The tree is gone now, cut down. But I still love this birch, and the fulfilling, lush airy memory of being fully immersed inside the swishing sounds of ten thousand small hanging leaves singing in the wind. From up there I could glimpse over the beech hedge and look much farther each time the branches would rustle in the wind, opening and closing windows to this larger world beyond.

Perhaps the reverberations from that old tree have made me reach out to a new one. A large pine bends over the path on a peninsula where I often go jogging. It is much, much older than my forty-seven years. Recently, I started paying more attention to it, feeling that I just have to stop and touch the bark before continuing. My eyes gravitate toward it and I greet it, like a rooted elder, before I get up close. I've sometimes brought my despair to this pine. And I like to imagine that this leaning tree has set down a few of its roots in me as well, that we share a Borderland where old dualistic and divisive splits no longer apply.

This capacity to bond with some aspect of the land seems never to completely leave us. The human soul loves to set down new roots in any nourishing place. It may happen while we're walking up a magical ridge, being greeted by a flower growing down in a canyon, watching a woman hang her laundry, or connecting with some feathered being singing on a twig.

## The Secret of the Despair Paradox

The term *biophilia* was first used by psychologist Erich Fromm to describe a human attraction to all that is alive and vital. It literally means "love of life" or "love of living systems." Ecologist Edward O. Wilson then used the term in the same sense when he suggested that *biophilia* describes "the connections that human beings subconsciously seek with the rest of life."[35] He proposed the possibility that the deep affiliations humans have with nature—the love of other species as our distant brothers and sisters—are rooted in our own human biology. And that the experience of this is a deep joy.[36]

Biophilia is not some abstract theory; it is the very sensuous joy of participating with clouds, winds, rain, ants, moose, birches, berries, birds, and even bears. Despair and depression do not necessarily block all this out forever. They can also reawaken and sensitize us to a keener wonder. When I'm stressed out or getting depressed, the best remedy is to get out on the land. Yes, we may know that it is all threatened, and perhaps forced into some suffering already. But it actually does not feel that way when I'm out wandering in some patch of land, wild scents spreading in the receptive air, and surrounded by plants large and small, including the deep-rooted old barked ones whose unflappable calm is contagious.

Our psyche is infinitively complex. Let us replace the psychological climate paradox with a "despair paradox": We can be deeply despairing, yet life can be really, really amazing. Enhancing our capacity for despair can also enhance our capacity for deep joy. Both reappear after numbing breaks up. As Derrick Jensen says: "I am full of rage, sorrow, joy, love, hate, despair, happiness, satisfaction, dissatisfaction, and a thousand other feelings. We are really fucked. Life is still really good."[37]

There's a hidden secret in the despair paradox: Going down to the depths of despair can also bring healing. Many environmental writers have talked about how respecting and cultivating grief evokes a deeper affection for the natural world. That pattern also corresponds well with most depth psychotherapies. The more we let death—and even the threats of extinction—into our soul, the more we can appreciate the current vibrant vitality of life in its many forms.[38] And we may even be transformed by it.

# SIXTEEN

# Climate Disruption as Symptom: What Is It Trying to Tell Us?

*The study of lives and the care of souls means above all a prolonged encounter with what destroys and is destroyed . . . We want to know what the symptom is saying about the soul and what the soul might be saying by means of it.*

—JAMES HILLMAN,
RE-VISIONING PSYCHOLOGY, *PP. 56–57*

When any of us decides to enter psychotherapy, it is mainly because we have a symptom that is bothering us. It may be depression, compulsive thoughts, addiction, eating disorders, pain, loss, or grief. The symptom feels alien, like an uninvited intruder that threatens to ruin our lives. We usually hate it, and want the therapist to remove it as quickly as possible. Yet by working it through, our life is often changed. Therefore, many psychotherapists refuse to be a gun for hire. They're not there primarily to terminate the symptom for the patient.[1]

If reimagined, the symptom has a large—though dark—power to transform. Hillman suggests we should be grateful for our symptoms; they often become guides for the soul-work of our lives. Gradually some sense of meaning and deepening may emerge from what used to be only seemingly disgraceful suffering.

Today, however, it is not just about human afflictions. Much of nature is sick, too. The animals we eat—domesticated or wild—contain PCBs, hormone disruptors, and radioactivity, the fish have too much mercury, the tap water tastes of chlorine and contains copper, some tap water from fracking areas catches fire, buildings have toxic indoor air, we breathe in dust and nitrogen oxides, and there's too much ozone near the ground but too little up in the stratosphere where it should be. The oceans are going acidic. Temperatures have gone off the charts. Forests are burning and fertile soil is getting depleted, eroded, or drought-stricken. The entire atmosphere seems to have caught some kind of fever, which is gradually infecting the oceans. The lament and complaints over nature's disruptions go on and on. The earth and the sky have now become full of seemingly meaningless symptoms and suffering.

Sigmund Freud initiated a line of analysis in which to examine culture with a pathological eye in the classic *Civilization and Its Discontents*.[2] Today, in the twenty-first century, it's time someone wrote the sequel *Nature and Its Discontents* with the same keen pathological eye. Hillman invites us to carry Freud's notion of therapeutic analysis beyond civilization to the whole of our experienced world.[3] Another way of not getting lost in the literal, meaningless despair over it all is turning to these calamities to regard them imaginatively as *symptoms and metaphors*. It may then be possible to find some sense of meaning in the fact that the air is getting sick or going crazy, too. Such eco-pathologies seem to want to tell us something that cannot be expressed by nature in any other way.

The conventional medical approach to malfunctions, however, is that symptoms should effectively and rapidly be removed by appropriate technology and procedures so that the ideal and healthy function is restored. For instance: Let's say you have a problem with an aching tooth. Solution: Fix cavities, or replace the tooth. Or let's say a car suddenly doesn't drive well on one side and has become difficult to steer. Solution: Trash the ruptured tire and replace it with a new one. What if your computer won't boot up? Solution: Reinstall the operating system.

Since such solutions-focused approaches have been so technologically successful over the previous centuries, most climate policies unquestioningly follow the same principles in treating the eco-pathologies:

- **Symptom:** Too much particle pollution from the smokestacks? **Solution:** Higher smokestacks, or install a smokestack filter.
- **Symptom:** Foul stench from growing heaps of garbage? **Solution:** Drive garbage to the landfill or incinerator.

- **Symptom:** Too much $CO_2$ in the atmosphere? **Solution:** Carbon capture and storage, which means pumping the $CO_2$ back underground. Or try geo-engineering by, for instance, fertilizing the algae in the ocean with iron.
- **Symptom:** People don't believe in climate change? **Solution:** Use peer messengers, positive framings, and generous nudging.

And so it goes. The same thought pattern is applied everywhere. We lift the train back onto the rails. We change things from "wrong" to "right," as defined by the "captain of the soul"—our supposedly rational, mastering ego.

Not all aspects of psychotherapy, systems thinking, and reflective medical practice are driven solely by solutions, though. And there's nothing wrong with practical solutions. But sometimes more respect—from *re-spect*, to look again—is needed to understand the symptom itself before we jump to implementing solutions. Depth and existential psychologies know that most of our key insights and fundamental knowledge derive from the state of soul *in extremis*: the sick, suffering, dying, and abnormal. We want to set ourselves straight from the sickness, but cannot.[4] The symptom speaks about something deeper, or beyond our solutions and conscious ideals.

Even such physical symptoms as a broken leg have more meaning than just fixing the bone and marrow. The meaning of the bodily symptom involves no longer being able to walk with your friends in the same way; you lose freedom, and might feel like you are broken, an outsider, or a victim. Or the accident may be framed as an unexpected opportunity for deep rest; a hole in time to read, rethink, and reorient your life in another direction. In other words, how you feel about your broken leg may be saying something about how you believe your life is going. The symptom—physical or psychological—speaks, if we let it.

Another key insight from psychotherapy and systems theory is that when we simply remove one symptom, without looking further at its context and wider causes, then we tend to quickly get another symptom from the same system. Taking a whole-systems view, we easily see that if you fix the tooth, but the patient continues to skip flossing and eat a lot of sugar, he'll quickly get yet another toothache. Or worse: For each symptom we fix, ten new ones appear. Suddenly there's diabetes and obesity, too.

This is not to say we shouldn't fix the tooth. Or that we should not put water-saving showerheads into our bathrooms, put efficient LED bulbs into lamps, put carbon capture and storage systems on the remaining fossil-fueled power plants. The situation requires we must do all these things, and more. But then, in parallel, we must also ask some vital questions: What is the symptom itself trying to express? Not: Who's to blame? But: What *story* is the symptom telling us? From *where* does it come? *Who* is speaking from it? Why does it show up *now*? What does it *want*? How is it trying to *change* me and us? Weaving an imaginative story around the symptoms can start a deeper understanding of our climate troubles.

## A Symptom Is Not Just a Problem to Be Fixed

Symptoms then are not just problems waiting for a fix. They are also a language. Problems can be seen as *em*blems of a larger picture. Psychotherapy has learned that through the human symptom the voice of the soul is speaking about its suffering.[5] And in an ecological symptom, maybe the voice of the earth is speaking. Studying these questions is the task of eco-psychology.[6] What do the ecological symptoms tell us about the state of our own minds and souls?

Eco-psychologists look at the dialogue between nature's way of thinking and humanity's ways of thinking. Why have we expressed our modern selves in such a way that our technological impacts have caused the entire air to get "sick," full of extreme weather events and apparently new symptoms? Maybe nature is trying to tell us something through its eco-pathologies. Maybe not just problems to be solved, but more like mysteries inside which we live? If so, what is it saying through the droughts, the depleting water tables, and the drying forests, as well as the immense floods and the intense heat waves?

In organizational theory, there is an important distinction between first-loop learning and second-loop learning.[7] First-loop learning refers to adjusting actions so we achieve the target that has been defined: Start brushing teeth to reduce cavities. The second loop includes questioning whether the defined target is the right one. Why is the focus only on fixing the tooth, and not on diet? First-loop organizational learning is to change the smokestack on a plant (make it higher, put a better filter in). The problem is "solved": The foul particle pollution is no longer to be seen or smelled. But second-loop questioning looks at why there is such a smokestack in the first place. Could we learn to redesign the products, the process, and the underlying values so there is no need for a smokestack at all? This is called whole-system redesign.[8]

Through second-loop learning, or simply deeper questioning, we quickly get into underlying values. What kind of values do the systems of highways, combustion engines, coal plants, airplanes, and agricultural monocultures express? Which of our values do high-rise buildings—and urban sprawl—actually embody? What are they emblems of? According to the founder of deep ecology, Norwegian philosopher Arne Næss, the inquiry into policy and lifestyles by such deeper questioning is at the core of eco-philosophy.[9] The climate symptoms are forcing us to re-spect, to re-look into the human values and our way of being in the world.[10]

The point is that developments in climate science itself are starting to shift traditional borders. Through the study of air and water ecosystems, we're increasingly revealing a natural world so thoroughly influenced by human activity that any clean cut between nature and culture can no longer be maintained. This forces us to rethink where we draw the system boundaries.

When global warming from fossil fuel emissions starts triggering feedback effects, such as melting permafrost and the loss of snow and ice that once reflected the sun's rays back into space, are we seeing nature at work, or human intervention, or both?[11] Similarly, the sinking water tables in dry regions might express something about the human thinking that guide our agroindustry and city planning. Or we know that tropical deforestation makes local temperatures go up and reduces rainfall, as in deforested parts of the Amazon. Is rainfall then separate from human-shaped landscapes? Or is changing rainfall *a reaction* to what we're doing to the forests and ocean temperatures? Rainfall and ice melt still retain their wildness, in the sense that we don't control their daily work. Yet they are reacting—in often unpredictable ways—to our interventions. Theirs is a feedback, not a natural course of events that would have occurred even without human influence. We are learning that the ecologies of Arctic ice and permafrost, smog and local rainfall are all highly sensitive to and react to the pervasive human influence. If these are re-actions, then there are also systemic messages here that may need to be listened to: Through its symptoms, we are called to discern what the natural world is telling us.

But hold on a minute—isn't this anthropomorphizing? This common counterargument might go like this: "Aren't you, Mr. Stoknes, making the assumption that there is a mind in nature that speaks back? Human technologies express human values, but floods and droughts and melting Arctic ice—all these things are purely natural phenomena. They have been happening for millions of years before humans walked on the earth and don't have any meaning. If you find message and meaning there, it is projection of your own mind. They are nothing but physical processes, driven by cause and effect. Aren't you stretching the envelope too far here? Leaving the safe border of sound science from parts 1 and 2 and now crossing over into mysticism and woo-woo?"

No.[12] The epistemological shift I'm suggesting is not wacky or anti-science, but a hermeneutical one—from only *explaining* the weather to also *understanding* what it is saying. Maybe there aren't only averages and causality, but also messages and meaning, something to be learned, from its responses. Meaning is not an exclusive human prerogative. Ask your cat who wants to be let out. A bear knows what it means when another bear false-charges it away from blueberries ripening in the late-summer sun. And a salmon knows something about the meaning of spring floods.

Thus there may be messages in the eco-pathologies. There is something to be learned when the canary stops singing, or all the baby fish in a bay die. They may be telling us something about how our actions influence the network of relations in which we live, despite being unable to speak and write English. Both explanation *and* understanding are necessary to arrive at a fuller picture.[13]

## What Climate Symptoms Say About Our Worldview

The concept of nature as a type of environment, something surrounding us that is distinct and distanced from us, was core to the development of modern Western individualism. And subsequently to environmentalism. But now the time may have come to let go of the word *environment* altogether, since there is a dualism or opposition built into it by definition: man versus environment, culture versus nature, subject versus object. We have learned to think of nature as something out there, in the forests, high in mountains, or deep in the oceans, somehow beyond and separate from us. Saying that "We have to care for our environment" leads us to think that we are here inside our houses and brains, while the environment is found beyond the fence and the street. In this view, we have to pull ourselves together to reach out and help this distant thing.

But the fragility of this opposition has become self-evident: When the fish disappear after overfishing, the crash in fish stocks results from the thinking that guided the fishing boats, not a distant environment. When the water tables sink and rivers in nature dry out, is our water consumption not involved? When the tundra starts to melt, allowing methane to bellow out from large blowholes, is that fully unrelated to human combustion? When the oceans heat and the floods become more deadly, science tells us we're implicated. And when the air I breathe is polluted and changed, this of course enters also my lungs, my very being.

It is not an opposition. Rather, we are fully *immersed* in nature. It permeates and envelops me as intimately as the air that holds my skin and dries its sweat, the food that melts on my tongue, and water that gushes down my throat. Being inside nature, or the environment, is very different from being separate and against it—or even advocating "for" it.

This new understanding applies not just to our intimate ecologies of water and airflows, but also to our worldview. From the previous century we have inherited what may be called a modern mind-set. In this mind-set, the earth is *boundless*; it and its oceans are regarded as huge, and the atmosphere as practically endless. It is also *robust*; it is governed by vast natural processes and laws, extremely resilient, and able to rebound from shocks or devastations and clean itself from waste and pollution. And finally, what happens to it is *determined by nature's processes*, which are considered so overwhelming that human impacts are negligible. In this modern mind-set, the main processes on Planet Earth are stable and predictable over millennia and millions of years. They run on their own cycles, influenced by forces like continental drift and occasional ice ages, wholly without human interference.

Now, however, the message from the air's symptoms seems to flatly contradict this mind-set. The earth no longer seems boundless; rather it seems more and more *finite*, if not smaller and smaller, in relation to the ever-growing human economy. Indeed, some resources are exhausted, and even more sinks of pollution are oversaturated. We are running out of planet to exploit.[14] It also no longer seems robust. Instead, it seems alarmingly fragile, capable of suffering from irreversible breakdowns. Waterways and winds seem incredibly responsive to human influences, and many delicate ecosystem balances are more precarious than imagined within the modern mind-set. Nor does it seem, any longer, to be determined by nature's processes. It is becoming apparent that humanity has become the most powerful force in shaping the face of the earth, and influencing the content of critical components in the air.

This amounts to an ongoing shift in our culture's foundational thinking, from a modern to an ecological mind-set.[15] From seeing our earth as boundless, robust, and determined by nature to a mind-set where it is seen as finite, fragile, and human-dominated. And the shift in mind-set is—to an increasing extent—driven by the climate disruptions themselves, particularly by large climate events like Hurricane Katrina and Superstorm Sandy. As long-lasting droughts spread and glaciers shrink, our old worldview is undermined. Just as stomach pains might alert us that we need to change our diet, we are receiving feedback in the form of eco-symptoms.

Despite the growing symptoms, our worldviews don't change easily. In public debates we can witness the resistance with which the old, modern worldview is withstanding and even actively combating this newer, more ecological worldview. The winds themselves, however, keep blowing and driving the changes, despite setbacks and social inertia. The next time you listen to a storm, you might hear the clouds shouting: "We've never been modern."[16] There is little doubt about who will prove the strongest in the end: the climate's violent symptoms or humanity's inert and defiant worldviews.

Our traditional Western individualistic ideas have made it difficult to incorporate aspects of family and nature into the very notion of self. But new ways to express this intuition of a more intimate relationship are being shaped through concepts of an ecological Self or an ecological identity.[17] This is somewhat akin to having a national, citizen, or ethnic identity. It refers to a sense of connection with the more-than-human world, particularly the plants, animals, and waterways of the places where we live. It expresses our connections to the land itself, and to the clouds, rains, and winds that move across it and nurture it. It also provides us with a recognition of similarity between ourselves and others.

The Alaskan author and anthropologist Richard Nelson writes beautifully about this: "There is nothing in me that is not of the earth, no split instance of separateness, no particle that disunites me from my surroundings. I am no less than the earth itself. The rivers run through my veins, the winds blow in and out with my breath, the soil makes my flesh, the sun's heat smolders inside me."[18]

First Nations people seem to always have a particularly keen sense of this extended sense of self. To know the land is also to better know yourself. Blackfeet elder Leroy Little Bear has put it this way: "You psychologists talk of an identity crisis. I'll tell you what an identity crisis is. It is when you don't know the land, and the land doesn't know you."[19]

Today, as the climate seems to go off in all directions and get weird and feverish, its symptoms lead to us to respect and recognize the land and its air again. They force us to start listening to its speech. The symptoms that in the beginning seem only alien and threatening to the ego consciousness eventually become the guide to understanding an expanded sense of self.

## What Climate Symptoms Say About Our Values

With this emerging shift in worldview and expanded sense of self comes a gradual shift in attitudes and underlying values. Some psychologists have started to measure the extent of this shift in individuals. They call it nature connectedness and have developed a scale to measure it.[20] Such connectedness refers to the degree to which individuals associate self with nature and believe that they are part of the more-than-human world.

Our values explain what we care about and why. Extensive research has identified three main sets of values: egoistic, altruistic, and biospheric. Egoistic values are focused on a limited sense of self, like achieving social power, wealth, and personal success. Altruistic values concern respect for other people, like family, friends, community, and broader humanity. Biospheric values focus on the health of beings like plants, animals, and trees, as well as mountains, rivers, or ecosystems. We all have each set of values, but they are weighted differently from person to person. Not surprisingly, research confirms that the more connected we are to nature, the stronger our biospheric values will be, and vice versa.[21]

Values can then explain *why* a person develops certain types of attitudes, which, if integrated into the self, in turn may shape behaviors over time. Researchers have found that when egoist-based values of power

and achievement dominate, people have less concern about how environmental damage affects other humans, children, future generations, and
more-than-human life. And when such self-enhancing values do promote
concern about ecological damage, this concern is limited to how such
damage might affect the self. The results from cross-cultural research are
quite clear: Expressing materialistic values is associated with less positive
attitudes toward nature and lower levels of biophilia.[22]

Our current economic system itself embodies values that fall into the
self-enhancing or egoistic categories, with its focus on wealth, personal
status, and success. What we fill our daily lives with—working for a salary,
purchasing and paying rent for houses and cars—automatically activates
and reactivates these values, so that the social and biospheric values get
relatively less mind-space. The stronghold that economic ideas and values
have on so many modern humans seems to be a basic foundation for rising
social marginalization and inequality as well as continued carbon pollution.[23] At the fundamental level, this is what is forcing the air to become
sick. And its symptoms return to us to ask us questions about the values
that dominate our culture and create the malaise in the first place. The
unhinged winds are hinting that our dominant values are crazy, at odds
with the air.

## Understanding the Limitations
## of Science and Psychology

Perhaps by now you assume that we can chalk the climate crisis up to
our modern worldview and our ego-based values, and that science and
psychology tell us all we need to know about how we arrived at our current
calamity. But that understanding would be incomplete. The symptoms of
the air seem to target some blind spots in science and in psychology, too.

Built into mainstream psychology's self-understanding as a discipline is
the idea that the psyche is somehow inside the individual, while the science
of ecology is about what happens outside the individual, in forests or oceans
or other natural spaces. Ecology and psychology, then, are two scientific
disciplines with very different borders, methods, and subject matters. This
used to be the "normal" state of affairs. But now—with the climate symptoms
—we've entered a post-normal stage: Human actions have influenced all

more-than-human interactions by shifting the temperature and weather patterns; through extensive land use; and through ocean acidification. Thus human thinking, action, and values can no longer be seen as wholly separate from more-than-human interactions.[24]

This exposes psychology's own denial. Conventional psychology has studied the human mind and emotion as if the climate, the air we're breathing, and the food we're eating don't really matter. The essentials for psychology lie only inside the brain; it has assumed that ecological concerns belong in some other department or silo. Thus psychology has had its own share of passive climate change denial. Until very recently climate change just wasn't considered something relevant to psychology or therapy.

Even now that it is becoming recognized as connected to modern stress, many continue to work as if the psyche could be fixed while the earth is abused and the air smeared. As if it is possible to sort out your childhood and your personal relationships, then go back to a lifestyle and job that kills off the ecology as we know it, and feel good about yourself. But it does not make much sense to keep working on your own self-development if your employment and society wipe out what therapy tries to rescue: a sense of belonging, meaning, safety, sanity, stability, and connectedness.

This is where ecological psychology or eco-psychology enters. Hillman encourages a therapeutic move from the mirror to the window.[25] If psycho-therapists and clients spent less time looking in the mirror and more looking through the window, they might better attend to the world. He invites psychologists to break away from staring into the mirror of introverted self-reflection to open out to the world. It is not just our own soul that individuates; it is the soul of the world, too.[26]

Like mainstream psychology, climate science has made the basic assumption that only human minds inside brains have some intelligence, have something to say. Ever since the Enlightenment, the air has been viewed as nothing more than inert dead gases. But in climate symptoms, we may eye the breakdown of a world whose depth and agency have been ignored by the prejudices of our climate science and psychology.

Simply put, climate symptoms can be seen as a calling to change our worldview, our values, and the framings of sciences that exclude subjectivity and intentionality from the world itself.

By becoming weird and sick, the air becomes visible and recognized; may even be revalued by humans. By delivering typhoons and droughts and scaring scientists with methane burps and blowholes, like reawakening an ancient dragon's breath,[27] the air is creating anxiety.[28] This angst breaks the naturalness of our habitual, conventional state of mind, a state Martin Heidegger names "das Man" or "the They," in which the air is nothing. But it now reawakens wonder: What is it up to? What is going to happen? What is it that we have to change?

Is the air, though, truly speaking to us? Are we willing to regard it as autonomous and communicative? And can its symptoms be seen as somehow purposeful—intelligent even?

# SEVENTEEN

# Re-Imagining Climate as the Living Air

*The wind blows wherever it pleases. You hear its sound, but you cannot tell where it comes from or where it is going. So it is with everyone born of the Spirit.*

—GOSPEL OF JOHN 3:8

It had been soggy and cold for days on end, our water bottles freezing at night, despite the fact that it was August. The higher reaches of Sarek National Park in Sweden can have arctic conditions even in summer. Still, three of us had had some great days of hiking from our base camp in the magnificent untouched alpine area with sharp peaks and glaciers, despite the heavy clouds, fog, rain, and sleet around us. On our last day there, we were roped in and walking up a glacier to ascend the highest peak, grandfather Mount Sarek himself. But with the heavy clouds hanging overhead, I thought: *What's the point? There's no chance of seeing anything at the top.* My older brother, however, is an utter peak maniac. He wants up no matter what. Just to have been there. So after traversing the glacier we started climbing the narrow shoulder of Mount Sarek, heading upward into the ever-thicker fog. The cliffs were barren and drizzling all over. We could hardly see forty feet up or down. Futile madness in the gray porridge.

Finally we spotted some whitening above our heads. We had no idea how far up we'd gotten. Then the air decided to show us some of its wizardry. As if we'd finally passed some kind of initiation test, it drew an amazingly sharp topline of the clouds. The peak of Mount Sarek extended up above them, like a huge shark fin swimming among endless white waves.

As we approached the top fifty feet, the wind we'd struggled against farther below was nowhere to be found. The stones at the peak were warm and dry. The brilliant sun we hadn't seen in a week was all around us. The smell of sun-dried black lichen wafted up from warm granite, and it was so warm we had to strip off all our clothes to the waist. We were stunned. The twenty-plus other peaks around us had all been magicked away. Nothing but our little patch of mountain was in sight. It was as if the world were suddenly made solely of bright white fluffy stuff, the deep blue above and this dark, moving shark fin we had to be careful not to fall off. My friend Mads wanted to take a picture with his mobile to send home to his wife. But there was no coverage of course. He snapped away, but when we came back he discovered the memory card had failed.

This active worlding,[1] this creation of a unique place, in that moment, remains now only as an embodied memory. I can still feel it. We never captured it on any flat, two-dimensional photo, nor could we have even if the mobile camera had worked, since it was so three-dimensional. What happened was that through surprising shifts, the air showed its powers to hide, create, reveal, and whisk out of sight again anything really, even massive seven-thousand-foot mountains.

Usually, the air is just the invisible background of our everyday life. We do not notice it. But when smog, smoke, fog, tornadoes, wind, snow, violent rain, or delicate dawn envelop us, we're reminded that the air is continuously bringing forth worlds inside which our bodies live. In certain Norwegian rural dialects, it's still common on a stormy day to hear, "*He* blows hard today." By personifying the air or the wind, we come to a closer relationship with it. Personifying is the heart's mode of knowing, says Hillman.[2] The invisible background may then become a clear foreground figure whose actions stand out more clearly.

The air is not an object. If anything, it is a grand subject, to which we are subjected. We are intimately connected to the air and others through our breathing. The air entering our own lungs and bloodstream has been carried by winds around the entire earth. With each breath we inhale quintillions of molecules of oxygen, which have been transpired not just by the trees in the street or garden, but also by the forests of the Amazon and particularly by phytoplankton in the oceans. We also breathe billions of molecules of $CO_2$ that were once emitted by volcanoes in Java or by pandas in China.

The environmental writer Paul Harrison, who argues for a scientific pantheism, states:

> Long before, this same oxygen and carbon were synthesized in the furnace of an ancient star which exploded as a nova or supernova, scattering its elements across space. Later they would condense again to help make our solar system, and were recycled through all life forms that have ever existed on earth. Every single breath you take contains the history of the galaxy.
>
> Every breath links us with every other human being that ever lived. Astronomer Harlow Shapley once calculated that our every breath contains more than 400,000 argon atoms that Gandhi breathed during his lifetime. Argon atoms are here from the conversations of the Buddha . . . and from recitations of classical poets. We have argon from the sighs and pledges of ancient lovers, and from the battle cries at Thermopylae. The simple act of breathing connects us in this intimate way with the plants and every person of the past of every race, religion and culture.[3]

But in the conventional, modern view the air is nothing but a mix of gases like nitrogen, oxygen, argon, and their now infamous cousins, carbon dioxide and methane. Stripped of wonder, it just passively fills empty space. In this reductionist mind-set the atmosphere is a complex system that behaves in unpredictable ways, but can still be approximated and appropriated by modeling, we believe. It may behave chaotically, but it is precluded from being seen as having *a mind of its own*. In the single vision of science, as the poet William Blake would say, it is not a person, it is not experienced as the Big Breath breathing us, thinking its cloud-thoughts as they shift toward the horizon.

## The Air's Poiesis: It Is Actively Bringing-into-Being

The three parts of this book—Thinking, Doing, and Being—are partly inspired by Aristotle's distinction of the three fundamental activities of psyche: *theoria* (thinking), *praxis* (doing), and *poiesis* (being). Poiesis means "knowing by creating." Something is brought into being, and the one who's creating is also being brought forth in the act of making. Anyone who has done any sort of improvisation has experienced *poiesis* in an exemplary way. In the creative process, new knowing simply arises. The dance arrives through the body's spontaneous movement; the jazz musician follows the flow of music to uncharted territory; the poet has to catch the poem before it flies away; the scientist loses herself for some time to the unknown she's studying; the writer's characters start writing their own dialogue; the actor improvises her way into expressing her character. All these are instances of knowing by bringing into being. The poiesis cannot be deliberatively planned, but must be developed by walking into and surrendering to the unknown.

We've seen how both conventional science and everyday opinion regards air not as something our psyche participates in, but as something out there, beyond us. We relate to it by observing it with measuring instruments, looking nearsightedly at computer models and graphic weather forecasts. We tend not to think much about the scents it carries, the breeze it brushes across our skin as we watch its shifting wind patterns. Nor is there any gratitude to the air for our very existence. We've distanced ourselves so much from the living air that only its revolt against us can break open the shell and dissolve this psychological distancing.

But what if we thought about climate disruption as the air's poiesis? The air is continually playful in the way it moves. It blows, moans, roars, whispers, and mumbles into the ears of each animal, carrying sounds from far away. It sings with the leaves and hosts infinite patterns of light, smell, sound, energy, and other waves. The air overflows with information, patterns, and voices, dynamically interacting with itself . . . and with us. What might happen if we stopped viewing the air as a blank canvas for carrying our radio waves or hosting our weather or bringing us the voice of our crying baby but—as Mount Sarek taught me—instead viewed it as the active imagination of the world? The fascinating thing is how it connects and intermingles with us and our psyche.

We can find hints in our language that this concept wasn't always overlooked. The very word *psyche* primarily harks back to breath, the big breath. It comes from the Greek *psukhē*, meaning "breath, life, soul." It is also related to the verb *psukhō*, "to blow, as the wind." Thus *psycho-logy* is really "to give an account (logos) of our life-breath (psyche)."

There is also *pneuma*, which in Stoic thought refers to the vital spirit, soul, or creative force of a person. In Greek *pneuma* literally is "wind, or that which is breathed or blown."[4]

The word *spirit* is from Latin *spiritus*, "the breath; to breathe." That's why we have words such as *inspire, respire, expire* in addition to *spiritual*; it's why we call certain moods *low* or *high spirits*. *Spirit world* thus refers to the world of the winds and the air.

In most of modern psychology, however, the psyche—commonly "the mind"—applies to what is assumed to happen only inside the skull, as if breath, body, and air do not participate. The psychologist Carl Jung put psyche at the center of his work. Yet he struggled to clarify his thoughts and concepts about it his entire life, particularly with how to limit or locate the psyche. Perhaps his most eloquent and unequivocal formulation is from his commentary on *The Secret of the Golden Flower*, a Taoist meditation guide, where he says: "The psyche is not any more inside us, than the sea is inside the fish."[5] This is a pretty potent statement: We are all swimming in a sea of psyche—it surrounds us on all sides, but we do not notice it.

If, as Jung also said, psyche creates reality every day[6]—and further, we are as much inside the psyche as the fish are inside the sea—then we're forced to rediscover psyche as breath and air. Every morning it is the air that lets light through to bring forth another "day." In this daily labor of creation the ground, the earth, the ocean, the waters, and the sun's fire give, of course, vital and necessary contributions. Then they all mix together in this sensitive and reflective medium, their meeting place, to create a "new day." Inside which we may go to work, and think nothing of it.

## Air and Psyche: We're Inside Air's Awareness

In phenomenology, consciousness is not inner observation of an object out there. Perception and consciousness arise in an active participation between whatever is sensed and the sensing body. Thus, our awareness is not just inside the brain. Becoming conscious is always a becoming aware of *something*, aware of what is around us; our bodily relation to friends, strangers, clouds, birds, squirrels, buildings, grasses, even dreams. Consciousness therefore involves a relation between our bodies and the bodies of other earthly beings. It is more encompassing than only the electromagnetic patterns encased inside our skulls. If the mind can't be separated from the body and the body can't be clearly separated from the world, then the mind can't be separated from the sensuous world.[7] As French philosopher Merleau-Ponty said: "There is no 'inner man.' Man is already in a world, and only in the world can we know ourselves."[8]

Eco-philosopher David Abram states that each land has its own psyche, its own style of breathing sentience. Hence to travel from Houston to San Francisco, or from Barcelona to Berlin, is also to voyage from one type of awareness to another. Maybe if we travel by foot or bike, this makes the shifts more evident than in a car: When the landscape gradually alters, mountains give way to foothills, and foothills become plains, we can feel the air change "as the moisture-laden atmosphere of the highlands, instilled with the breath of cool, granitic cliffs and the exhalations of roots and matted needles, opens onto the dry wind whirling across the flatlands."[9] In tandem with the shifting terrain and atmosphere, the accents of the locals, farmers and shopkeepers, also transform. Their way of noticing, naming, and speaking follows the character of the land and its air.

The South African Bushman-friend and explorer, Laurens van der Post, has a vivid passage that describes the continuum of awareness and air beautifully:

> From the ridge where we seated ourselves we had an immense view of the desert. In that light it looked in terms of earth what the sea is in terms of water—without permanent form and without end . . . When one has lived as close to nature for as long as we had done, one is not tempted to commit the metropolitan error of assuming that the sun rises and sets, the day burns out and the night falls, in a world outside oneself. These are great and reciprocal events . . . I was convinced that, just as the evening was happening in us, so were we in it, and the music of our participation in a single over-whelming event was flowing through us.[10]

The sunset—air set alight by the sinking sun—is a living field that we reciprocate with. Therefore the sunset, colors and all, is not just subjectively happening inside our heads. Rather, we're inside it. And *it* vibrates in *us*. The air and I, just like Laurens van der Post and the evening desert sunset, participate in each other to co-create this vast now.

But is this not complete madness? Are we out of our minds? Is this not a surrender to an animistic, severely romantic, or even psychotic world-view, a highly non-scientific approach? How can such a 180-degree turn be understood in a precise way?

Let me try with a story. Everywhere from the dusty desert of South Africa to the freezing snow in Norway, the air does its artful patterning in tune with the land. One day while I was writing this book, the silent workings of the air made a spectacular display of its creativity. During a few cold, starry winter nights, the air had been playing with condensing droplets, making ice art from the night's dew. The air had brought forth a miniature white forest of branching ice crystals, mimicking the larger boreal forest stretching out thousands of miles from where I now stood, and topping the snow on the ground for as far as I could see. One to two inches tall, each mini crystal tree joined with a million other crystal trees to form angle after angle, branch after branch. Together, they created a new layer, an extra cushion on top of all the snow, increasing the intimate and intricate interactions between air and earth.

I bent forward, nose all the way to the ground, to better study the myriad details inside the miniature trees. After a long while, when the cold had crept into my knees, I walked farther into this huge artwork. Then, to add to this overwhelming wonder—and as if to supersede itself—the air started to play by drawing the cloud curtains open to the winter sun's low afternoon rays. Some crystals were attached to grass poking out from the snow layer, and a soft breeze began to move them gently—softly enough that the crystals did not break off and fall. But it was still enough to make them play more vividly with the light. Soon the whole opening in the forest turned into a giant light-circus, easily outperforming any spectacle of Christmas lights I've ever seen. Millions of crystals glittered in the low, cold sun. Some switched on and off according to the soft movements of the grass. Depending on where I stood or how I moved, they winked and twin-kled. *Hello*, they seemed to say, now here, now there. As rays of sunshine jumped from grass stalk to grass stalk I followed the smooth and rapid movement among them. To my eyes, they were playing among themselves, sending light back and forth.

I walk in this living, pulsating, radiating field, a huge image exqui-sitely crafted in the stillness of the night and then made animatedly alive by the interplay of wind and winter sun in the day. My consciousness is not located outside, observing this image. I am carnally immersed in it. Participating with my entire body and particularly intensely through this peculiar, watery, double sensitivity to the world of light that we call *eyes*. The air fills the space completely, sets the scene; over some hundred years it lets the trees grow, conjures forth the magnificent crystals with effortless ease during a frosty night, lets the sun in at the appropriate moment, arranges all beings in their right seats, holds around them with a perfectly pressured embrace, nudges the grass to move, and in general hosts the entire spectacle, it seems, out of pure joy.

Is there awareness in the air? Are the grass and the crystals part of it? And did they mind a two-legged participating in this performance? Would there have even been a show without me? I tend to think my presence was welcome. Consider the phrase *Esse est percipi*; to be is to be perceived, as Bishop Berkeley claimed. But spectacles like I witnessed that day surely don't depend on the presence of a human perceiver. There are sufficiently sensitive beings all through the forest field to make up an enthusiastic audience—if one is even required.

We have long ascribed awareness only to what we already regard as animate, and even among the animate we argue about what is fully conscious—what can communicate or act with intention, what can feel and express emotion. While cultures of long ago did not hesitate to ascribe emotion or intention to other animals, to plants, or even to forces of nature, we do. So if I were to suggest that all this art was "made" because the air felt like playing with ice and water crystals the night before, I might be viewed as being slightly off my rocker. But let's dig a little deeper.

To understand the air as an active, creative force, we need to forget all the narrowly human aspects that we might ascribe to creative acts. I can't say, for instance, the air is an artist that created a spectacular art show, collecting the tickets, and expect it to be taken as anything but an analogy. But is there a way of viewing the air as something as creative and alive as we are, albeit in an altogether different way? And if we viewed the air that way, might we be more responsive to its way of communicating?

Perhaps the interplay of forces that created my light show, combined with my perception of them, is what used to be known as *anima mundi* in the Western history of ideas, the "world soul" according to Plato, the Neoplatonists, and the Stoics. The world soul is the flip side of modern dualism, where thinking is exclusively done by human subjects. In worldviews such as Plato's, imagination does not lie solely in the human brain. Imagination lives in the greater world, within each being and outside of it. It is an exchange, and it plays out in the world around us, hosted by the air. Thus, my awareness is participating in and co-creating the larger field of awareness in the forest.

This seems rather close to the portrait Carl Jung began to paint of our "psyche" as the sea we swim in. It also seems close to what the philosopher Martin Heidegger called Dasein, or Being-in-the-world.[11] Both thinkers struggled to articulate and communicate their deep understanding of these fundamentals. In one graceful—if intricate—formulation, Heidegger expresses his insight this way:

> Man is never first and foremost on the hither side of the world, as a "subject," whether this is taken as "I" or "We." Nor is he ever simply a mere subject which simultaneously is related to objects, so that his essence lies in the subject–object relation. Rather, before all this, man in his essence is ek-sistent into the openness of Being, into the open region that lights the "between" within which a "relation" of subject to object can "be."[12]

I'd like to repeat Heidegger's last phrase: the open region that *lights* the between *within* which a *relation* can be. This openness of Being is, quite simply, the air. The air holds the space within which all our relationships play out. It is the prerequisite for my meeting with anything and anyone. This open holding it provides is not just passive, but also co-creative, since it brings us all into presence. What he may be getting at is that Being is more a verb than a noun, that the world is more appropriately seen as somehow participatory and aware.

Today, in the midst of climate disruptions, it may be high time that we recover and remember this sense of the air as, in its own way, alive and sentient. That fact that we have forgotten the air, replacing it with the abstract concept of gases, also explains our irreverence for what we now call the atmosphere. We have forgotten that the air is a sacred, intelligently creative being. Our societies are unthinkingly letting it receive all our wastes. Filling the air with exhaust, fumes, smoke, dust, chemical compounds, airplanes, and noise. Is it any surprise that the air, having tolerated this behavior for too long, now is violently shifting into another mode altogether? Why should it put up with our folly any longer?

## Historic Roots to a Sense of the Living Air

If this view of the air as a living force with a complex agency seems strange today, it is not in any way new.

At the very beginning of our Western culture's most basic text and creation story, we meet the air and wind. There are two key quotes: "And the Spirit of God was hovering over the face of the waters." The original Hebrew word for the "Spirit of God" comes from the Hebrew phrase *ruach elohim*, a great wind. And then the second: "And the Lord God formed man of the dust of the ground, and breathed into his nostrils the breath of life; and man became a living soul."[13]

So wind and breathing, air and humans, are connected and closely related to the most sacred at the roots of Western culture. If the Spirit of God is identical to the great wind blowing, how come the cultures that grew out of sky religions have attacked the sky with emissions and pollution? Both modern Christianity and Islam claim there is only one true God in heaven. But removed from the concretely sensible sky. God is no longer recognized in the great wind hovering over the waters.

With God removed from the perceivable heavens, and the sense of the sacred removed from the air, the full-scale attack could begin. We attack the sky with fire set in great forests, smokestacks, millions of exhaust pipes, combustion, huge gas leaks, and more. Our technologies are titanic; think of giant coal mines, refineries, supertankers, and huge power plants. Mythology is a fantasy that reveals some level of truth, and with that in mind we can ask ourselves: Can we look at severe weather and other climate disruption as a sort of reenactment of the revolt of the ancient Greek Titans? Is our unleashing of titanic technologies a new uprising against the sky gods that have maintained a balanced cosmos over the last millennia?

Various religions have long entertained an image of going to heaven for the afterlife. If you're a believer, that may be the case. But heaven may not be exactly as some have imagined—a transcendental realm far above, a permanent celestial spa, outside time and place, somewhere beyond the earth altogether. Another way of imagining it is more immanent: Our little breath eventually returns to the Big Breath. When I draw my last breath, the little wind that whirls inside my lungs and blood permanently returns to the larger wind around me and the larger whirls of clouds and sky. I surrender fully to It. There is no contradiction of ordinary science here.

We only imagine the same transition differently, more vitally. The air is no longer just seen to be an empty dead envelope of invisible gases. It is where our life-breath comes from and what it returns to.

If we start imagining the air differently, we may also discover that what used to be called heaven is not somewhere we have to die to travel to. We may be in heaven already, and continuously so. Perhaps 'going to heaven' is better viewed as a metaphor for reawakening to the beauty of this invisible moving mystery that we are inside at this very moment.[14]

If Christians, for instance, were to widen the lens on their views of heaven, they might also be able to reimagine Jesus's main message: "Turn to God and change the way you think and act, because the kingdom of heaven is near!"[15] As near as the air, you might say. Some scientists would be alarmed at the analogy, but what climate science tells us today isn't that different: Turn to reason and better ways of thinking and acting because the climate changes are imminent.

In a conversation not long ago, a devout Christian said she wasn't worried about climate change because God granted humans dominion over other creatures, we've acted accordingly, and if it seemed like we were in deep trouble now we shouldn't worry about it. God will take care of things for us. Would her views change if she were to broaden her lens?

Conjecture aside, though, it is clear that the wind, air, and heavens have been seen as sacred in most native cultures, long before modernity's chemical and physical definitions of air started to displace all other understandings.

## Indigenous Traditions and the Holy Wind

Among most indigenous culture and First Nations we find a strong recognition of the air and the wind. In the Navajo worldview, for instance, the "Holy Wind" holds a central importance.[16] The basic association between life and breath has been developed into a comprehensive theory of life, motion, thought, speech, and behavior. Navajos conceive of Wind as a unitary being, there being only one Wind, which is the source of all life, movement, and behavior.

Yet—since the Wind appears differently at specific times and places, it has been given many names. The four principal Winds are White Wind, Blue Wind, Yellow Wind, and Dark Wind. They may correspond to the white of Dawn, the blue of the Sky, the yellow of Twilight, and the dark of Night. Whirlwinds are also important, turning either clockwise or counterclockwise.[17]

To the individual person, the winds called Wind's Child, or the Little Wind, are vital because they are small enough to enter into bodily orifices and inhabit the body. Once connected with individuals, these winds can teach and guide them if they really listen. For instance, some of these Messenger Winds are sent by the inner forms of the sacred mountains, and serve as agents for the leaders or chiefs. The Little Winds are sent primarily by the Holy Ones in the cardinal directions to influence human thought and conduct.[18] The Wind with which we are born and the Winds from the cardinal directions will take care of us and work to protect us from harmful influences. Each of us can be shown the right way to behave and be warned of dangers.

To the Navajo, then, the air and its winds have properties that European civilization has mainly seen as the inner mind or "psyche." But if these are really powers of the air—both entering our body as well as blowing around us—then this mind or psyche is not really ours, is not an only-human possession. "Rather, mind as Wind is a property of the encompassing world," writes David Abram, "in which humans—like all other beings—participate. One's individual awareness . . . is simply that part of the enveloping Air that circulates within, through, and around one's particular body."[19]

Simply put: Our own life-breath is a small wind that interacts with the large wind around us. There is a continuum between both my breath and the wind, both my psyche and the awareness of wind and air. Abram points out that this has immediate impacts on ourselves: "Any undue harm that befalls the land is readily felt within the awareness of all who dwell within that land. And thus the health, balance and well-being of each person is inseparable from the health and well-being of the enveloping earthly terrain."[20]

## The Air and I: Practical Implications?

All this philosophy about the air and awareness may seem speculative and academic. What difference does it make if awareness is inside the skull or continuous with the air?

It's no remarkable insight at this time to realize that humanity is now on the way to disrupting biodiversity and the climate. We've been pretty busy with that for some centuries already. What is remarkable is that some humans are beginning to wake up, as from a centuries-long slumber, to rediscover our intimate relationships to air, sea, forest, ourselves, and one another.

My point is that this is all about reawakening to wonder. Seeing the air with new eyes. There is magic in the air—if we attend to it. This is not extra-reality magic or superstition. There is no need for extra-sensory perception or parapsychology or appropriation from other cultures' spiritual traditions. For us, it is about reawakening to the surprising creativity that the air is performing in front of our very, but weary, eyes. The shifting clouds, the endless colors of dawn, rainbows, and fog: All this is a form of everyday magic. This is the magic of the real, not the supernatural. But to really notice it, our own perception must shift a little. This shift to recognizing the living air changes everything, but not by changing any-thing.

Take, for instance, our attitude toward clouds. They're not just pretty, or heavy. As heat shields, living duvets for the planet, and water transport, they're critical not just to our hydropower, but also to our survival. Without them, the sun would obliterate life.

And they draw our imagination. We see elephants and twisted horse heads and all manner of images in their billowing, woolly whiteness as they shape-shift above us. They wander their own ways, awakening our never-ending wonder. This continuous creating and re-creating, the free play of the skies, is also happening in our awareness as we contemplate clouds moving, or the way a butterfly bounces along, or the way morning mist slides through the trees.

This understanding of being as an active creating process, as *poiesis*, has particular relevance to everyone who works with creative processes, because this means that creation is continuous. Many of us have tended to view creation as what happened to earth five billion years ago; after that, it has been up to the physical laws to mechanistically run the show. But in reality, each present moment is pregnant with creation. We're co-creating the world in any now.

When I participate in elaborating an "image"[21]—in science as a new theory or discovery, in writing, ritual, or song, with brush or without, in forest or in city, playing an instrument or not—I participate, too, in the Big Breath's continual bringing forth of images. The air is the imagi-nation, the nation of images we're residents in. Within it, I participate in a kind of reciprocal imaginative interaction with other things and beings. This is also participating in a larger ongoing creation. This view bestows a vast sky over every creative act, however humble or seemingly insignificant the images we work with.

If it's only up to me or us to act, and counter climate destruction, then hopelessness can and will sneak in. We may feel little reason for optimism. We're so small. If, however, we're teaming up with the air, and its continuous worlding, then we may feel we act in resonance with a larger context, flow. By aligning myself with this larger wind, I gain deeper connectedness: I am the wind talking through my throat (therefore I'd better be careful with my speech). I am part of air caring for itself (am I helping or hurting it?). I am part of a billion-year-old being, starting to implement the necessary adjustments.

I am the sky thinking itself in me, claims phenomenologist Merleau-Ponty. He writes: "As I contemplate the blue of the sky . . . I abandon myself to it and plunge into this mystery, it 'thinks itself within me,' I am the sky itself as it is drawn together and unified, and as it begins to exist for itself; my consciousness is saturated with this limitless blue." Merleau-Ponty's "I" in this quote is no longer distanced or opposed to the sky. It expresses the lived body, my flesh, intimately participating with the flesh and breath of the world. And finding a tranquil joy in it.

The shift to realizing that we're deeply embedded in this living air is the deep reframing for climate communication. I hardly see anyone in today's climate discourse speaking of air as something enchanted, beautiful, and sacred—except maybe some on the fringes of society, often dismissed as crazies. But unless we bring in this ancient aspect, the awe for the air, we'll lose something central and powerful—a sense of the sacred that can be critical for motivating us in the highly needed transformation, that makes climate into something very near, no longer just distant, abstract, technical, chemical. For all the amazing things the air does, and daily, it deserves such a sense, no matter how we choose to name it or understand it; Heaven, Holy Wind, Great Spirit, openness of Being, the Imagination, or simply the Air. Otherwise, why bother for a dead blob of gas?

So walk into a vibrant forest and reimagine the exchange and fine-tuned communication everywhere around you, the complex interaction of birdcalls, dew, pollen and pheromones, dense patterns of light, sound, smell, and more all hosted and held and mixed by the air. Or the next time you walk into a strong wind, play with it, say hello by leaning into it and letting it move you. Or just feel the delight when the gentle afternoon breeze rustles your hair. We're held. As the children of this "great spirit," hovering over land and waters, we might even feel loved—not only by friends and family, but also by this magnificent being of clarity and light.

Therefore, first and last, there is gratitude in the act of reimagining air. It is gratitude for the life it always has and always will nourish.

# EIGHTEEN

# It's Hopeless and I'll Give It My All

*Rows and floes of angel hair*
*and ice-cream castles in the air,*
*I've looked at clouds from both sides now*
*From up and down, and still somehow*
*It's clouds' illusions I recall*
*I really don't know clouds at all*

—JONI MITCHELL, "BOTH SIDES NOW"

*Hope* is such a muddled word: "We must have hope!" "There is light at the end of the tunnel!" Why do climate books and speeches have to end on a note of hope?

I've asked the question "Are you hopeful?" to a series of climate communication researchers. Everyone I ask starts off with a laugh or smile, since they can neither answer yes nor no. One answered, "Cautiously optimistic." Another said, "That depends on the season; this spring I was severely negative, now I'm more hopeful."

You can't find a politician, however, in or running for office, who is a pessimist. They all say they were born die-hard optimists. As if positive thoughts are wired into their genetic code, they are all brim-full of hope. They tell us there will be progress on every front and more jobs for all. Provided they get elected, of course. They are acutely aware that hell doesn't sell. Votes are not won by spreading gloom. Gloom is only helpful if it can be blamed on their opponent's policies. So optimistic hope has to win the day. At least on the surface of things.

But optimism by itself may easily become overblown and spill over into wishful thinking. Insisting on positive thinking, simple solutions, and bright futures may devolve into a type of denial by itself.

I have filled up part 2, Doing, with positive strategies, since these connect better to human needs for glow and flow. Yet sticking only with the glossy view despite better knowledge can lead to self-deception, and the deception of others. History is strewn with failed rosy visions and crushed dreams. A classic tragic case of hope's slippery slope can be found in former British prime minister Neville Chamberlain's "Peace for Our Time" statement, in which he relayed his positive feelings that peace could be brokered across all of Europe. He spoke those words less than a year before Europe plunged into World War II. Hope also failed those who tried to take on an optimistic outlook from within the Warsaw Ghetto during the Nazi occupation of Poland.

On the other hand, from the scientific and academic camps, pessimism comes in bucket loads. We hear from them that "Time is running out to meet the two-degrees target." And: "Vast glaciers in West Antarctica seem to be locked in an irreversible thaw linked to global warming that may push up sea levels for centuries."[1] Exact but depressive analyses are in excessive oversupply, far outweighing the demand for them.

It is true that most megatrends for the natural world point in the wrong directions: climate gas emissions, water and food use, forest cover, ocean acidity, ecosystem biodiversity. It is easy to point out that our societies respond too slowly to counteract the escalating problems, and that the fossil infrastructure and reserves already in the pipeline to be extracted doom us to shoot far over the safe limits of global warming, into the realm of weird, dangerous, and utterly unknown effects. Clever, intelligent pessimists seem to have all the best data on their side.

One of the things I learned during my almost two decades of strategic scenario consulting and teaching is that both pure optimism and pure pessimism make very poor scenarios. They have very little to do with any real future. If you think that everything will be bright, *or everything will be dark*, you are guaranteed to get the future wrong. Such optimism and pessimism do not by themselves make complete stories of futures. Rather they are psychological perspectives, simplifying stances, which should be inherently mixed into all futures thinking. Thus, all plausible long-term scenarios should be fundamentally different and each should include both positive and negative elements, different shades of light and shadow. Otherwise we're back to one of the most fundamentalist stories in our culture: the monotheistic outcomes of either heaven or hell, absolute optimism or absolute pessimism, salvation or damnation. Which dooms us to the too-common manic-depressive cycles of activism.

I've also learned that since the long-term future is fundamentally uncertain, the future really must be thought about in the plural form: as futures. We simply do not know how it will play out toward the end of the century. Therefore we need multiple stories, at least three or four at a time that are all equally plausible though deeply different in many ways. Even if most of the long-term futures look grim, like a muddling-through at best, we still don't know enough to be cocksure. Sometimes, things turn out differently and even better than hardly anybody can predict.

Therefore I'm not a pessimist or an optimist. Neither am I a realist, as if the gray middle between the two is the most plausible outcome. Optimism and pessimism are tools we can apply when considering the wildly different futures lurking beyond the horizon. And they are best used in parallel, like the left and right eye. It's not an either–or, but a both–and. The optimism of part 2 on Doing is needed, but so too is the feeling of despair and game-over for the stable climate and the many fellow species that we used to know. Part 2 is incomplete without part 3 on Being. Thoughtless hope clings to optimism to repress despair and avoid the hard, no-frills look at the factual disgrace. Thoughtful pessimism only comes up with ever more reasons why our outlook is grim and harsh.

At a seminar, poet Gary Snyder was once asked, "Why bother to save the planet?" He replied with a grin: "Because it's a matter of character and a matter of style!"[2] What I really like about his answer is that it doesn't attempt to found our actions on some plausibility calculation of success or failure, or on a dualistic ethics—the good fight against evil. Rather Snyder points to our calling and to aesthetics, both realms of the soul, of being who we are.[3] This grounds our long-term actions in something more substantial than the expectations of a quick and successful outcome of our efforts. For sure, early wins and successes are welcome and wanted. We just can't make our long-term efforts dependent on them.

We humans are not necessarily destined by our genes to self-destruct by short-termism. There's a cultural layer on top of it. What parts I and II have shown is in summary the following: If we believe that other people don't care about this distant climate issue, if it is framed as cost and sacrifice, if it's expensive and inconvenient to act green, if we think inside the story of economic efficiency and growth, and, finally, if all signals show that we're heading toward hell, then many of us will shut off our minds and act short term.

But at the same time, if we believe that others really do care and the climate is framed as insurance and health for ourselves, it's easy and simple to act green, and we share stories of opportunities for jobs and the good life. And, finally, if we can see that society is making steps in the right direction—if some or all of these solutions are shaping the situations around us—then most humans will act for the long term. There are thus grounds for hope, but we must separate hope from bland optimism, and distinguish among the different varieties of hope.

## Varieties of Hope

When the massive destruction of land, ocean, air, and livelihoods is documented, many start feeling hopeless and helpless. Others insist that we must keep up the hope, and that the only thing to fear is fear itself. But what do we actually mean when we use the word *hope*?

One version of hope is based on *passive optimism*: 'Oh, things will turn out well. Technology will fix it for us. Nature has made climate change before.' As a personal life-stance, optimism may have much that speaks for it: "Don't worry, be happy. Most of my worries never materialized. Maybe a little bit of nature is dying, but it will take care of itself." I'll call this type of hope Pollyanna hope or passive hope. It is an outlook where—if you only think positively—all is sweetness and light. Since the world turns out well anyway, there is no reason to worry and work; we can wait for rewards to ripple down our way.

Another type of hope is much more *actively optimistic*: "The best way to predict the future is to invent it. We'll make it happen. There is no end to human creativity and ingenuity; where there is a will, there will be a way." This is also a *Yes we can!* and *Just do it!* attitude. This type of optimism says that the likelihood of a good outcome depends on the magnitude and acumen of our effort. It may be a fight, but one we're going to win: "We're strong and going to do what it takes . . ." This type is a heroic hope.

To defend optimism-based hope, both the passive Pollyanna and the active, heroic hope, you have to keep believing in the likelihood of good outcomes. There must be a good reason to be optimistic—either because things end well all by themselves or because we make them come out well. In optimism we get attached to the likelihood of certain favorable outcomes. But if the outcomes threaten to turn sour and dark, this type of optimism easily crumbles into pessimism.

There are many examples of our inability to do what it takes to solve the climate crisis, and they are repeated often in the media. Recent book titles are revealing: Naomi Oreskes published *The Collapse of Western Civilization: A View from the Future*. Clive Hamilton's book is called *Requiem for a Species*. *Scientific American* called theirs *Lights Out—How It All Ends*. Jared Diamond's is known as simply *Collapse*. Thom Hartmann's is titled *The Last Hours of Humanity: Warming the World to Extinction*. Guy McPherson's book is called *Going Dark*. Derrick Jensen's 640-pager is *Endgame*. IPCC's lyric contribution to the genre is "RCP8.5."

Optimism has—scientifically—a weak case. But if optimism is unfeasible and glaringly utopian, is pessimism then inescapable, and hopelessness inevitable?

No. There is a third way—or more—of hope. This way involves a true skepticism, one that embraces the sense of not-knowing, that accepts the extent of the unknown unknowns. Nobody knows enough to be an absolutely convinced pessimist. Things may look bad, sure, optimism may seem rationally impossible to me today, but that doesn't necessitate flicking the switch over to pessimism. That happens only if we think in terms of dualistic opposites (or some grayish in-between). We don't have to be attached to either end of that line.

This opens up to a view of the future beyond optimism or pessimism. It is a form of skepticism that also comes in a passive and active version:[4]

The *passive skepticism* leads to a type of stoic hope: "We'll weather the storm." "We don't know what's coming. But whatever comes, we'll take it. No matter, we'll stand our ground." "After the tempest, we'll rebuild. We're resilient. We've been through hard times before, and can handle them again." This is a version of hope that's sturdy and hardy, not clinging to optimism, but still making no proactive effort to dream or influence the future.

The *active skepticism* is somewhat more demanding to describe. It goes along the lines of "There's no reason to be optimistic, but we're going for it anyway." Or: "Our situation is desperate and at the same time hopeful." Desperate because of the dynamics now unfolding, yet hopeful because we remain unflappable. In this brand of hope, I'm not attached to optimism or to pessimism. I call this stance a *grounded hope*. It's grounded in our being, in our character and calling, not in some expected outcome. The future is fundamentally uncertain and complex. Therefore it is open to the imagination and always possible to influence in some way. So yes, it's hopeless *and* we're going all-in. The active skeptic gives up the attachment to optimistic hope and simply does what seems called for. There is a deep freedom in that. The dream sometimes glimmers like a silver thread, and that's all I need to keep walking. I don't need to believe that things will end well in order to act. The walking and the doing are their own reward.

## The Revitalization of Citizen Community

My son and I were out walking on a quiet city street in Oslo, toward the old garden behind King's Castle. *Wham*—a huge, sharp blast came out of nowhere. I assumed something really big had fallen off a truck just around the corner. But as we kept walking, we saw no truck. Well, mysteries abound, and we continued on to my office. I had important stuff to do, while my son pulled out his iPad. Just an hour or so later, I happened to browse an online news service. There I saw breaking news of an immense terrorist explosion that had nearly brought down the prime minister's office. Lots of dead, but no suspects yet. I suddenly got it—that sharp blast had been an explosion less than a mile away.

Then the really bad news started: unconfirmed reports of shootings at a youth camp outside Oslo. As if one shock were not enough. I left for home with my son, and from home we watched the unraveling of Norwegian innocence: The man, a nationalist blond beast, clad in police uniform, was walking around on the island of Utoya, executing teenager after teenager. Systematically shooting them through the head, three to five shots each, sometimes singing. Their crime? Working for an inclusive society, belonging to a party that welcomed—in his opinion—too many foreigners. It was a well-prepared attack—not on an ideology, but on the caring, intelligent future generation of our small country. The next days were full of news, unfolding stories of horror beyond belief. The number of dead grew and grew, eventually reaching sixty-nine, with sixty-six badly wounded. On top of that there was the toll of the city blast.

Two days later, a few people used Facebook to set up a rally in solidarity with the victims. This is where cities reach their full potential: The rally went viral. After residents had spent days huddled individually in front of their TV screens, it was time to take to the streets. With only one day's notice, two hundred thousand citizens of Oslo turned out. It was impossible to walk the rally route, since the entire area was so filled with people. With the center chock-full, everyone gathered around the city hall. Most brought roses. Walking quietly, astounded. Gentle despite the throng. It was not just the survivors and families of the victims or the political party that had been hit. We had all been hit. That afternoon the sense of community, of the city's soul, was palpable; the air filled with roses, quickly assembled music, hushed voices, shared despair, and incredible determination.

It moves me to tears each time I recollect it: "If one man can show so much hate, think how much love we could show, standing together," said one survivor from that island. Cities and citizens gathered and moved toward a shared purpose: the revival of generosity and community after a storm of terror. The prime minister underlined from the stage that the proper answer to the violence was not surveillance but "more democracy, more openness, but not naïveté." I had lived in Oslo for more than twenty years, but I had never before witnessed so many Norwegians—typically reserved—meet one another in such an inclusive way. No doubt about it, the capacity for opening up to reciprocity is there, as demonstrated in many other cities, too, such as New York after the 9-11 blasts and Sandy, the solidarity with New Orleans after Katrina, and in Berlin after the fall of the wall.

This sense of shared openness and commitment to care now needs to be extended to the more-than-human world.

## Expanding Our Sense of Self and the Reversal of Agency

To the modern mind, individual action is thought to start from within willpower—the singular "I" that, if motivated, sets things in motion toward a goal. In this view we are all rational actors, improving our situation by making choice after choice. It takes willpower and muscle to enforce the decisions of the thinking I onto the world. This I, this Western ego, sees itself as an isolated entity even if related to other individuals. It must draw nourishment for change from its own inner—but brittle—determination.

The previous chapter elaborated a different view of the human–world relation—one where we act in concert with the wider world around us, in ways we may not entirely understand and that remove the boundaries among ourselves and the air we breathe. In this view, we don't act on behalf of our air or on behalf of our ecology. We act *with* them because we are part of them and they are part of us. So, when taking action for more well-being with less destruction, we act in-spired by the air. Thus we re-spire and take part in the earth caring for itself. There's nothing supernatural in this. It is just breathing, seen from a broader view.

Breathing is more than an exchange of chemical elements between two separate entities. It is a sensual immersion. Pay attention, and we can notice how our breathing connects us to a larger being, inside which our lives play themselves out. Noticing our breathing—something meditative practices from all over the world have endorsed for millennia—brings our awareness more in touch with this expanded awareness. The meditative traditions have so far focused a lot on our own breathing, but much less on the air.

The old, worn adage of *Save the planet* can now be reframed. It is no longer just us working to save a planet that is out there. The planet itself is at work—the larger biota and air rebalancing itself, through a little help from you and me.

There is a grounding, hopeful message here: Ecosystems are working together with us to protect life. The air is bringing its disrupted state and inevitable responses unmistakably to our attention. The earth has hosted life for billions of years, delicately balancing conditions so that life has survived. Who knows what tricks the deep ocean has in store? Who truly knows what the clouds are up to?

The earth has experienced dramatic shifts in greenhouse gas concentrations before, from methane burps to giant volcanic eruptions. For some colossal, prehistoric events, the bounceback has come in geologic time, not the lifetime that you and I are hoping to see change within. But on smaller time scales, the earth's self-restorative capabilities are on display daily. It is time for humanity to heal from its self-imposed separation and fully participate in the earth's self-healing processes.[5]

## Flowing with the Winds of Change

Twenty years ago I had a dream that has stuck with me since. I was heading up some snowcapped mountains on a snowy road to spend time at a cabin on the other side. But I couldn't get there. A friend had given me a keycard to the ski lift that could take me there, but I had left it down below, in my car, and now I was somehow in a bus, but the bus had gone too far. Off the bus and walking back, I met Picasso along the road. He showed me a painting of a simple Picasso-style tree bending in strong winds. "It's titled *Activist*," he said.

Stronger winds are blowing now, and not only the hurricanes and Hadley cells. The disruptions of weather and air are also driving societal, cultural, and inner shifts. Some long for a cabin outside it all, beyond the hills in undisturbed nature. In the dream, this move to get out of it all is frustrated. When I try to opt out with ski lifts, cars, and buses, they bring me back to the road. Where surprisingly Picasso—the great imagist—is walking. His picture shows me an Activist, he says. But the real activist is not the tree that is pictured. It's the invisible wind who is the activator, bringing the tree into movement.

The dream left me with a feeling that it's not me, as an upright being, who is the activist and the doer, but the wind. I can let myself be blown through by this larger flow—my responses shaped, bent, and sustained by my way of standing in this vibrant current.

If engaging with climate is not a matter of winning, but more a matter of character and style, then the making of art, story, and literature also becomes part of our responses. In the novel *The Plague*, the French existential philosopher and author Albert Camus describes an isolated town struck by disease and how people respond to this threat to their lives and way of life. The novel follows Dr. Rieux as he struggles with his dying patients and relays the citizens' response to the plague. Many live as before, just denying it. Others fall apart in angst or apathy. In Dr. Rieux, Camus portrays a person choosing an engaged life even though the world around him is disintegrating. The world seems absurd and devoid of meaning. A faceless force is threatening to wipe out all of them with brutal suffering. Ethics philosopher Clive Hamilton sees a strong parallel between Camus's plot and the climate fight. Hamilton points to Dr. Rieux as an example of how "one should start to move forward, in the dark, feeling one's way and trying to do good."[6] Similar to what I called grounded hope above, his is an amalgamation of hardheaded pessimism with a die-hard optimism. Not a bright, heroic fight, but a firm refusal to capitulate and a steadfast commitment to keep going, despite hopeless odds.

Yet even Albert Camus is not consistently absurdist. Toward of the end *The Plague*, Dr. Rieux and his friend Tarrou, in a key scene, sit on the terrace of a house and gaze over the sea far into the horizon. They decide to go for a swim in the ocean, which they had long been locked out from. The sea was "deep-piled velvet, supple and sleek as a creature of the wild."[7] Just before Rieux enters the water, he is possessed by a "strange happiness," a feeling that is shared by Tarrou. There is a soothing image of Rieux floating motionless on his back in the sea, gazing up at the sky. The air opens up to reveal the stars and the moon. He draws a deep breath. When Tarrou joins him, the friends swim side by side. After this participation in the air and the sea, they are renewed and can return to the city for the final stages of the plague.

Rieux and Tarrou are not just recharging their batteries by having a picnic. Camus is drawing on deeper archetypes, consciously or not, and his scene is a superb depiction of our lived participation in the world—the coming into presence with sea and air and moon that sustains us. The plague didn't kill them all.

We are not separate from the climate. We are intimately inside the living air. That is the ground of the grounded hope. Every breath holds the potential to re-mind us of it, nourish us in the endless struggle against failing odds. But for too long we have not noticed what sustains us. We forgot wind, the morphing clouds, the rooted ones, and the deep seas. We thought we were self-contained and could only understand the air from a place somehow outside of it; we thought we would have to draw all motivation from inside ourselves. But climate disruptions open the portal to remembering the wonder of being-here, of being held and inspired by the living earth. There is an opportunity for trust and rest in its flow—like an owl hanging silently in the night sky, or an eagle lifted effortlessly above a mountain, or two friends floating in the sea under the stars—before harder strokes are needed again.

There are too many good reasons why we humans resist the many sad facts of climate disruption, the global weirding. It finally boils down to the question, Why bother? That one question reveals a simple fact: The most fundamental obstacles to averting dangerous climate disruption are not mainly physical or technological or even institutional; they have to do with how we align our thinking and doing with our being. This missing alignment shows clearly in the current lack of courage, determination, and imagination to carry through the necessary actions. But these human capacities are, luckily, as renewable as the wind and the sunshine are. By coming into presence, opening our chest and belly, the air fills us anew.

# —NOTES—

## Introduction

1. Hamilton, *Do You Trust Scientists About the Environment?* See figure 6.
2. Quantifying the exact level of consensus in a complex field with a plurality of definitions is a difficult task, which is why the Intergovernmental Panel on Climate Change (IPCC) and National Climate Assessments exist. The number 97.4 (as a percent of climate scientists) is from Doran and Zimmerman, "Examining the Scientific Consensus on Climate Change." The range 97 to 98 percent is from Anderegg et al., "Expert Credibility in Climate Change." After studying 11,944 published papers on climate change, Cook et al., "Quantifying the Consensus on Anthropogenic Global Warming in the Scientific Literature," conclude that "Among abstracts expressing a position on AGW, 97.1% endorsed the consensus position that humans are causing global warming." One contrarian article in a non-climate science journal that is contradicting the consensus claim is Legates et al., "Climate Consensus and 'Misinformation.'" The scientific consensus might recently have grown even stronger than 97 percent. James Powell (2014) has finished another such investigation, this time looking at peer-reviewed articles published in 2013: "For 2013, there are 1,911 articles [search terms global warming and global climate change] + 8,974 [climate change] = 10,885, 2 of which are rejections, about 1 in 5,440." Thus only 0.02 percent of articles published in 2013 explicitly reject human-made global warming. A scientific consensus doesn't get stronger than that.
3. See, for instance, Pew Research Center, *Climate Change and Financial Instability Seen as Top Global Threats.* This report gives the following percentages of respondents saying "global climate change is a major threat to their countries": United States 40 percent, Canada 54 percent, and Europe 54 percent, while Latin America is 65 percent. See figure 1.3. In the Pew report "More Say," only 45 percent of US residents answer yes to the question "Do scientists agree earth is getting warmer because of human activity?", p. 3. There are many different methods and surveys to measure concern, but no standardized global tests; see Pidgeon, "Public Understanding Of, and Attitudes To, Climate Change." More about this in chapter 1.
4. This is based on model runs he has completed with a team at the Norwegian Business School since publishing *2052*.
5. Many now blame the Intergovernmental Panel on Climate Change for being too careful and not sufficiently strident in its formulations. See Brysse et al., "Climate Change Prediction"; Scherer, "Climate Science Predictions Prove Too Conservative."
6. See chapter 1 for details. The international comparison is in: Nielsen Company, *Sustainable Efforts and Environmental Concerns Around the World.*
7. We've heard that repeated yearly since 1987. World Commission on Environment and Development, *Our Common Future.*

8. Painter, *Climate Change in the Media.*
9. Aakre and Hovi, "Emission Trading."
10. The US Environmental Protection Agency rules are a special case: According to a poll, 70 percent of US citizens support carbon regulation, yet action is blocked in Congress (as of June 2014) by Republicans. This is a case of dysfunctional representative democracy, gerrymandering, and the climate issue being given lower priority than other issues.
11. The European Union, in October 2014, pressed through stronger goals for 2030: a 40 percent binding reduction for all countries. In the United States, a Yale study shows that as of 2014, 37 percent of voters say they are willing to sign a pledge to vote only for political candidates who share their views on global warming. If this increased to 47 to 57 percent, it might be sufficient—over time—to shift the debate, despite gerrymandering, low voter turnout, and congressional dysfunction. See Leiserowitz et al., *Politics & Global Warming, Spring 2014.* Yet we also know that voter power often or usually loses out in influence on public policy compared with the influence of economic elites and the stands of organized interest groups. See US analysis in Gilens and Page, "Testing Theories of American Politics." Thus, there is no automatic voter victory in representative democracies, even if it can—with persistent effort—be made to work.

## 1. The Psychological Climate Paradox

1. As documented by IPCC assessment reports 3, 4, and 5. IPCC Working Group 1, *Climate Change 2013: The Physical Science Basis.*
2. G8+5 Academies, *G8+5 Academies' Joint Statement*, p. 5.
3. Muller, "The Conversion of a Climate-Change Skeptic."
4. See the booklet by the Royal Society and National Academy of Sciences, *Climate Change: Evidence and Causes*, or the full IPCC *Fifth Assessment Report: Synthesis Report.*
5. Lakoff, *Don't Think of an Elephant!*, p. 17.
6. Nisbet and Scheufele, "What's Next for Science Communication?"; Nisbet, "Communicating Climate Change"; Sturgis and Allum, "Science in Society"; Whitmarsh, O'Neill, and Lorenzoni, *Engaging the Public with Climate Change*; Wolf and Moser, "Individual Understandings, Perceptions, and Engagement with Climate Change."
7. Monbiot, "Climate Change Enlightenment Was Fun While It Lasted."
8. See Swift, "Americans Again Pick Environment Over Economic Growth; Hellevik, *Ipsos MMI's Survey "Norsk Monitor"*; Nielsen Company, "Sustainability Survey"; Pew Research Center, *Climate Change and Financial Instability Seen as Top Global Threats*; Nisbet and Myers, "The Polls—Trends."
9. Mooney, "The Strange Relationship Between Global Warming Denial and . . . Speaking English."
10. Pew Research Center, *Climate Change and Financial Instability Seen as Top Global Threats*, p. 1.

11. Dr. Maxwell T. Boykoff in Nielsen Company, "Sustainability Survey," based on a survey of more than twenty-five thousand Internet respondents in fifty-one countries.

12. Different studies unfortunately use different phrases and different sets of topics when measuring people's top policy priorities. But climate clearly gets very low overall ratings. Gallup in 2014 shows "climate change" as 14th of 15 (see Riffkin, "Climate Change Not a Top Worry in US"). Pew Research Center *Public Priorities* in 2012 shows "global warming" as 22nd out of 22, while Pew Research Center *Deficit Reduction Declines* in 2014 shows "global warming" as 19th out of 20. See also Brulle, Carmichael, and Jenkins, "Shifting Public Opinion on Climate Change" for a more in-depth discussion.

## 2. "Climate Is the New Marx": The Many Faces of Skepticism and Denial

1. Stoknes and Hermansen, *Lær Av Fremtiden*.

2. For a classic work on skepticisim, see Naess, *Scepticism*. Naess takes Pyrrho's "abstaining from final judgement" as the essence of the skeptic position.

3. The classic work on this is Kuhn, *The Structure of Scientific Revolutions*.

4. One example of a contrarian skeptic position is expressed by climatologist Judith Curry, when she writes: "By their nature, unknown unknowns are unquantifiable; they represent the deeper uncertainties that beset all scientific endeavors. By deep, I do not mean to imply that they are necessarily large. In this review I hope to show that the scope for revolutions in our understanding is limited. Nevertheless, refinement through the continual evolution of our understanding can only come if we accept that our understanding is incomplete. Unknown unknowns will only come to light with continued, diligent and sometimes imaginative investigation of the data and metadata." In Curry, "Uncertainty in Sea Surface Temperature Measurements and Data Sets."

5. I do not claim this list of examples to be complete, exhaustive, or statistically representative, but rather selected typical illustrations of texts and reasoning. They are all true in the sense of being quotes and referenced, and thus they are valid as examples of what passes for climate denial.

6. Høydal, "Hvem Bryr Seg."

7. See US House of Representatives Committee on Energy and Commerce Minority Staff, *The Anti-Environment Record of the US House of Representatives 112th Congress*; repeated in Michaels, "Why Hasn't the Earth Warmed in Nearly 15 Years?" He is also a director at the Cato Institute, a right-wing think tank, which has a history of climate science denial. For more on Michaels and other public misinformers, see Theel, "Patrick Michaels."

8. Bell, "Global Warming Alarm"; Bell, "The Feverish Hunt for Evidence of a Man-Made Global Warming Crisis."

9. Bell, *Climate of Corruption*.

10. Meador, "Kentucky Lawmakers Attack Climate Change Science in Discussion on Carbon Regulations."

11. All the following quotes are from the US House of Representatives Committee on Energy and Commerce Minority Staff, *The Anti-Environment Record of the US House of Representatives 112th Congress.*

12. Brulle, "Institutionalizing Delay"; Cook, "Attacks on Scientific Consensus on Climate Change Mirror Tactics of Tobacco Industry."

13. Inhofe, *The Science of Climate Change.*

14. Committee for a Constructive Tomorrow (CFACT), "Release: Call to Suspend Climate Treaty Negotiations at CFACT Press Conference."

15. http://wattsupwiththat.com/2014/07/29/temperature-analysis-of-5-datasets-shows-the-great-pause-has-endured-for-13-years-4-months/#more-113693.

16. *Washington Times*, "Chilling Climate-Change News."

17. Moran, "Alarm on Global Warming Just a Load of Hot Air."

18. Amos, "Arctic Summers Ice-Free 'by 2013.'"

19. This phenomenon is often called false balance in the literature. See, for instance, Painter, *Climate Change in the Media*, chapter 4.

20. From a reader's comment to Revkin, "On 'Unburnable Carbon' and the Specter of a 'Carbon Bubble.'"

21. From www.wattsupwiththat.com, July 29, 2014, at 3:32 PM.

22. Sometimes commentary gets really dark, particularly in the United States, with phone calls and hate mail threatening murder. See, for instance, http://desmogblog.com/2014/03/25/exclusive-climate-change-philosopher-target-abusive-hate-campaign.

23. Lewandowsky, Oberauer, and Gignac, "NASA Faked the Moon Landing—Therefore, (Climate) Science Is a Hoax"; Lewandowsky et al., "Recursive Fury."

24. Freud, "On Negation" (1925) and "Fetishism" (1927), in J. Strachey (ed.), *The Standard Edition of the Complete Psychological Works of Sigmund Freud*, vol. 21 (London: Hogarth Press), pp. 149–57.

25. Cohen, *States of Denial*, p. 32. See also Cohen, "Discussion," in Weintrobe (ed.), *Engaging with Climate Change*, chapter 4, "Climate Change in a Perverse Culture," p. 75.

26. Cohen, *States of Denial*, p. 103.

27. Norgaard, *Living in Denial*, loc. 177 in Kindle version.

28. Ibid., loc. 1100.

29. Cohen, *States of Denial*, p. 251.

30. Hoofnagle, "About," *Denialism Blog.*

31. Diethelm and McKee, "Denialism."

32. Hoofnagle, "About," *Denialism Blog.*

33. Lifton, "The Climate Swerve."

34. More on this in chapter 4. See also Hernes, *Hot Topic—Cold Comfort*; Kahan, "Climate Science Communication and the Measurement Problem."

35. The original article was Hardin's "The Tragedy of the Commons" from 1968. This has since been critiqued, for instance by Ostrom, "Revisiting the Commons." Still, many political scientists and economists refer to it as a valid barrier.

36. Of course, many governments do not even work for all citizens of their own country, but rather for a smaller, select group of citizens, including their own network. Such corruption is widespread, particularly in the energy and petroleum sectors.

37. *The Economist*, "Unburnable Fuel." Except that when pushed by activist shareholders, oil companies including ExxonMobil, Shell, and Statoil state that they don't see political regulations affecting their asset values. See Hope, "What the Fossil Fuel Industry Thinks of the 'Carbon Bubble.'"

38. See Leaton, *Unburnable Carbon*; www.carbontracker.org; McKibben, "Global Warming's Terrifying New Math."

39. Norgaard, *Living in Denial*, loc. 205.

40. Hernes, *Hot Topic—Cold Comfort*.

41. Hulme, *Why We Disagree About Climate Change*.

42. Keukens and van Voren, "Coercion in Psychiatry."

43. In March 2013, Pew Research Center asked thirty-seven thousand people in thirty-nine countries: "Do you think that global climate change is a major threat, a minor threat or not a threat to [survey country]?" The percentage of people who responded "no threat" or "minor threat" is typically 30 to 60 for most wealthy Western countries, but lower for Japan, South Korea, Brazil, and Greece. Pew Research Center, *Climate Change and Financial Instability Seen as Top Global Threats*, table Q11g. For Norway, see Austgulen, *Nordmenns Holdninger Til Klimaendringer, Medier Og Politikk*.

44. Foucault, *History of Madness*.

45. Freud illustrated this tension between creativity and destruction with the help of Greek mythology: Eros (as the life force and love) and Thanathos (the death wish and pleasure of killing).

46. Brulle, "Institutionalizing Delay"; Cook, "Attacks on Scientific Consensus on Climate Change Mirror Tactics of Tobacco Industry"; Oreskes and Conway, *Merchants of Doubt*.

47. Manne, "How Vested Interests Defeated Climate Science."

## 3. The Human Animal, as Seen by Evolutionary Psychology

1. Griskevicius, Cantú, and Vugt, "The Evolutionary Bases for Sustainable Behavior." See also Ornstein and Ehrlich, *New World New Mind*, chapter 7.

2. Griskevicius and Vugt, "Let's Use Evolution to Turn Us Green."

3. Yet indigenous knowledge can teach modern cultures fundamental ways to pay more attention to and live more intertwined with the nature–culture nexus. See Wildcat, *Red Alert!*

4. Jared Diamond's version of the story, in *Collapse*, has also been criticized, for instance by Hunt and Lippo in *The Statues That Walked*, in which they state that introduction of rats, disease, and slave raids were responsible. In a reply Diamond says, "The islanders did inadvertently destroy the environmental underpinnings of their society. They did so, not because they were especially evil or deprived of foresight, but because they were ordinary people, living in a fragile environment, and subject to the usual human problems of clashes between group interests, clashes between individual and group interests, selfishness, and limited ability to predict the future." More of Diamond's reply can be found here: http://www.marklynas.org/2011/09/the-myths-of-easter-island-jared-diamond-responds, accessed October 1, 2014.

5. McCarthy's novel *The Road* was in 2009 made into a dystopian movie.
6. Wilson, Daly, and Gordon, "The Evolved Psychological Apparatus of Human Decision-Making Is One Source of Environmental Problems."
7. Dawkins, *The Selfish Gene*.
8. Leiserowitz et al., *Climate Change in the American Mind, April 2014*. October 2014 figures from Leiserowitz, pers. comm., in press for January 2015.
9. Barclay, "Trustworthiness and Competitive Altruism Can Also Solve the 'Tragedy of the Commons'"; Griskevicius, Cantú, and Vugt, "The Evolutionary Bases for Sustainable Behavior."
10. Dr. Stockmann, in Henrik Ibsen, *An Enemy of the People* (1882), act 4.
11. Griskevicius, Cantú, and Vugt, "The Evolutionary Bases for Sustainable Behavior," p. 122.
12. Cialdini, Reno, and Kallgren, "A Focus Theory of Normative Conduct"; Cialdini, "Crafting Normative Messages to Protect the Environment."
13. The classic groupthink case analysis is John F. Kennedy's eight- to ten-person team of presidential advisers responsible for the failed Bay of Pigs invasion of Cuba, analyzed in Irving Janis, *Groupthink*.
14. One example that actually mentions Janis's research is contrarian journalist Paul MacRae's piece on the website wattsupwiththat.com: MacRae, "Why Climate Science Is a Textbook Example of Groupthink." There is a list of psychological shortcomings in MacRae's article too long to be discussed here. The main counterpoint is that his analysis doesn't check Janis's six to eight critical criteria that distinguish groupthink in teams, and thus it is not a textbook example as McRae superficially claims.
15. John Heywood, a sixteenth-century collector of proverbs, recorded this version in his ambitiously titled *A dialogue conteinyng the number in effect of all the prouerbes in the Englishe tongue*, 1546.
16. Griskevicius, Cantú, and Vugt, "The Evolutionary Bases for Sustainable Behavior," p. 123.
17. From Gallup research; see Riffkin, "Climate Change Not a Top Worry in US." APA writes in a report: "In 2009, as many as 75 to 80 percent of US respondents said that climate change is an important issue, yet they placed it twentieth out of twenty compared with other issues." American Psychological Association's Task Force on the Interface Between Psychology and Global Climate Change, "Psychology and Global Climate Change," p. 127.
18. A little caveat in case somebody wants to arrest my use of examples: I do not claim that California wildfires and Beijing smog are evidence for climate change. They are extreme (with regard to our cultures' experiences) weather-related events emerging from recent human–environment interaction. Each single event is more like weather than like climate, which is a statistical average over time. I use the examples rather as visualizations of, and metaphors for, the scientific prediction that climate change will—over time—increase the frequency and/or severity of such extreme events.
19. Gilbert, "If Only Gay Sex Caused Global Warming."

## 4. How Climate Facts and Risks Are Perceived: Cognitive Psychology

1. Harvey, *Cognition, Social Behavior, and the Environment*; Fischhoff, "Hot Air: The Psychology of CO-Induced Climatic Change."
2. NOAA National Climatic Data Center (NCDC), "Global Analysis—Annual."
3. American Psychological Association's Task Force on the Interface Between Psychology and Global Climate Change, "Psychology and Global Climate Change."
4. Ritter, "Climate Psychology 101."
5. Donner and McDaniels, "The Influence of National Temperature Fluctuations on Opinions About Climate Change in the US Since 1990."
6. Leiserowitz, "American Risk Perceptions: Is Climate Change Dangerous?"; Spence, Poortinga, and Pidgeon, "The Psychological Distance of Climate Change"; Spence and Pidgeon, "Framing and Communicating Climate Change."
7. Newell and Pitman, "The Psychology of Global Warming," p. 1007.
8. Kahneman, *Thinking, Fast and Slow*. System 1 is fast, intuitive, and unconscious; 2 is slow, deliberate, and conscious.
9. Weber, "Experience-Based and Description-Based Perceptions of Long-Term Risk."
10. Weber et al., "Heuristics and Constructed Beliefs in Climate Change Perception."
11. Ibid.
12. Kahn, "Hurricane Sandy Hasn't Shifted Climate Narrative."
13. This example was given by a participant, who presented himself as being from the farmers' union, at a seminar where I gave a keynote address. Norwegian Business School, November 29, 2013.
14. Nielsen Company, *Sustainable Efforts and Environmental Concerns Around the World*.
15. Hunt, "Interpreting the Nielsen Study . . ."
16. Newell and Pitman, "The Psychology of Global Warming"; Sterman, "Risk Communication on Climate." Part 2 of this book is dedicated to this topic.
17. Bensinger and Vartabedian, "Toyota to Fix 'Very Dangerous' Gas Pedal Defects."
18. For overview on the risks of dying, see Britt, "The Odds of Dying"; National Science Council, "Odds of Dying."
19. Intergovernmental Panel on Climate Change (IPCC) Working Group 1, *Climate Change 2013: The Physical Science Basis: Fifth Assessment Report, Summary for Policymakers*, D.3.
20. Loewenstein et al., "Risk as Feelings."
21. Ibid.; *Economist*, "Daily Chart: Danger of Death!"
22. Ropeik, *Risk*.
23. See also Marshall, *Don't Even Think About It*, chapter 10.
24. Loewenstein et al., "Risk as Feelings"; Marshall, *How to Engage Your Community and Communicate About Climate Change*, chapter 11; Morgan, *Risk Communication*; Weber, "Experience-Based and Description-Based Perceptions of Long-Term Risk."
25. Sinaceur, Heath, and Cole, "Emotional and Deliberative Reactions to a Public Crisis."
26. Lakoff, *Don't Think of an Elephant!*
27. Ibid., p. 3.
28. Leiserowitz et al., *What's in a Name? Global Warming vs. Climate Change*.
29. Li, Johnson, and Zaval, "Local Warming."

30. Bojinski et al., "The Concept of Essential Climate Variables in Support of Climate Research, Applications, and Policy"; NOAA, "NOAA's Ten Signs of a Warming World."
31. Luntz, *The Environment: A Cleaner Safer, Healthier America.*
32. "Global Warming and Climate Change Myths," *Skeptical Science*, www.skeptical-science.com, accessed June 5, 2014.
33. Leiserowitz et al., *What's in a Name? Global Warming vs. Climate Change.*
34. In the Pliocene, three to five million years ago, the earth had around the same $CO_2$ levels as it does today. The global average temperatures were 3 or 4 degrees Celsius (5.4 to 7.2 degrees Fahrenheit) higher than today's and as much as 10 degrees C (18 degrees F) warmer at the poles. Sea level ranged between five and forty meters (16 to 131 feet) higher than today. http://keelingcurve.ucsd.edu/what-does-400-ppm-look -like. See also Csank et al., "Estimates of Arctic Land Surface Temperatures During the Early Pliocene from Two Novel Proxies."
35. Previous warmings at the end of an ice age, however, happened very fast—in a decade or so; one example is the warming after the Younger Dryas. Also, rapid climate changes have been more deadly than asteroid impacts in the earth's past; see Howard, "Rapid Climate Changes More Deadly than Asteroid Impacts in Earth's Past—Study Shows."
36. Lakoff, *Don't Think of an Elephant!*
37. Kallbekken, Sælen, and Hermansen, "Bridging the Energy Efficiency Gap."
38. Norton, "Constructing 'Climategate' and Tracking Chatter in an Age of Web n.0," p. 6.
39. Ibid.
40. *EcoWatch*, "Police Close 'Climategate' Investigation into Hacked Emails with Mystery Unsolved."
41. Kahneman, Knetsch, and Thaler, "Experimental Tests of the Endowment Effect and the Coase Theorem"; Thaler, *Nudge*; Tversky and Kahneman, "Rational Choice and the Framing of Decisions."
42. Tversky and Kahneman, "Loss Aversion in Riskless Choice."
43. Ibid.; Spence and Pidgeon, "Framing and Communicating Climate Change."
44. Hardisty, Johnson, and Weber, "A Dirty Word or a Dirty World?"

## 5. What Others Are Saying: Social Psychology

1. Lindzey and Aronson, *Handbook of Social Psychology.*
2. Intergovernmental Panel on Climate Change (IPCC) Working Group 1, *Climate Change 2013: The Physical Science Basis: Fifth Assessment Report, Summary for Policymakers.*
3. Jasanoff, "A New Climate for Society."
4. Benestad, "A Failure in Communicating the Impact of New Findings," italics mine.
5. Lindzey and Aronson, *Handbook of Social Psychology.*
6. Overviews of climate attitude polling are given by: Brulle, Carmichael, and Jenkins, "Shifting Public Opinion on Climate Change"; Pidgeon, "Public Understanding Of, and Attitudes To, Climate Change"; Nisbet and Myers, "The Polls Trends";

Wolf and Moser, "Individual Understandings, Perceptions, and Engagement with Climate Change."

7. Leiserowitz, Maibach, and Roser-Renouf, *Saving Energy at Home and on the Road*.
8. Leiserowitz et al., *Climate Change in the American Mind, April 2014*.
9. How much people actually know depends to an extent on how we measure it, and which countries we are speaking of. See Leiserowitz, Smith, and Marlon, *Americans' Knowledge of Climate Change*; Austgulen and Stø, "Klimaskepsis i Norge."
10. Festinger, Riecken, and Schachter, *When Prophecy Fails*.
11. Festinger and Carlsmith, "Cognitive Consequences of Forced Compliance"; Festinger, *A Theory of Cognitive Dissonance*.
12. Hamilton and Kasser, "Psychological Adaptation to the Threats and Stresses of a Four Degree World"; Stoll-Kleemann, O'Riordan, and Jaeger, "The Psychology of Denial Concerning Climate Mitigation Measures: Evidence from Swiss Focus Groups"; Whitmarsh, "Behavioural Responses to Climate Change."
13. Here I'm simply pointing to how we tend to think about the issue. I'm not making judgments about accuracy, or indicating that climate ultimately really is an individual issue with individual responsibility.
14. Klein, "How Politics Makes Us Stupid."
15. Kahan, "Making Climate-Science Communication Evidence-Based—All the Way Down," p. 11.
16. See Hernes, *Hot Topic—Cold Comfort*; Marshall, *Don't Even Think About It*, chapter 6.
17. Franklin, *The Autobiography of Benjamin Franklin*.

## 6. The Roots of Denial: The Psychology of Identity

1. People who are skeptical about climate change tend to be older, white, male, and politically conservative. Austgulen and Stø, "Klimaskepsis I Norge"; Kahan et al., "Culture and Identity-Protective Cognition"; McCright and Dunlap, "Cool Dudes"; Poortinga et al., "Uncertain Climate."
2. See McCright and Dunlap, "The Politicization of Climate," figure 4, p. 176; see also Jacques, Dunlap, and Freeman, "The Organisation of Denial," and McCright and Dunlap, "Anti-Reflexivity," for further discussion.
3. Brulle, Carmichael, and Jenkins, "Shifting Public Opinion on Climate Change"; Brulle, "Institutionalizing Delay."
4. Hamilton, "Why We Resist the Truth About Climate Change"; McCright and Dunlap, "Cool Dudes"; Oreskes and Conway, *Merchants of Doubt*.
5. Kahneman, *Thinking, Fast and Slow*; Newell and Pitman, "The Psychology of Global Warming"; van der Linden, "Exploring Beliefs About Bottled Water and Intentions to Reduce Consumption"; Whitmarsh, "Scepticism and Uncertainty About Climate Change."
6. Feldman et al., "The Mutual Reinforcement of Media Selectivity and Effects."
7. Kahan et al., "Culture and Identity-Protective Cognition"; Kahan, Jenkins-Smith, and Braman, "Cultural Cognition of Scientific Consensus," p. 149; Kahan et al., "The Tragedy of the Risk-Perception Commons"; Kahan, "Making Climate-Science

Communication Evidence-Based—All the Way Down"; Kahan, "Fixing the Communications Failure."

8. Kahan, Jenkins-Smith, and Braman, "Cultural Cognition of Scientific Consensus," p. 157.
9. Kahan, "Climate Science Communication and the Measurement Problem."
10. Ibid.; Mooney, "Conservatives Don't Deny Climate Science Because They're Ignorant. They Deny It Because of Who They Are."
11. Lewandowsky et al., "Recursive Fury"; Lewandowsky, Gignac, and Vaughan, "The Pivotal Role of Perceived Scientific Consensus in Acceptance of Science"; McCright and Dunlap, "Anti-Reflexivity"; McCright and Dunlap, "The Politicization of Climate Change," Poortinga et al., "Uncertain Climate."
12. Mooney, "Conspiracy Theorists Are More Likely to Doubt Climate Science."
13. National Public Radio (NPR), "A Christian Climate Scientist's Mission to Convert Nonbelievers."
14. Hoofnagle, "What Is at the Root of Denial? A Must Read from Chris Mooney in *Mother Jones*."
15. Diethelm and McKee, "Denialism: What Is It and How Should Scientists Respond?"; Smith and Leiserowitz, "The Rise of Global Warming Skepticism."
16. See Lewandowsky et al., "Recursive Fury," p. 36; see also Marshall, *Don't Even Think About It*, chapter 8. A full-blown such narrative fills an entire book by Ball, *The Deliberate Corruption of Climate Science*.
17. McKewon, "The Use of Neoliberal Think Tank Fantasy Themes."
18. Kahan, "Fixing the Communications Failure."
19. Hamilton, "Why We Resist the Truth About Climate Change."
20. Kahan, "Climate Science Communication and the Measurement Problem."
21. See Weintrobe, *Engaging with Climate Change*, chapter 3.
22. Brulle, "Institutionalizing Delay"; Oreskes and Conway, *Merchants of Doubt*.
23. Eliot, *Four Quartets*.
24. Cohen, *States of Denial*, 265.
25. Ibid., p. 259.
26. Bagley, "Q&A: How a SuperPAC on a Shoestring Is Taking on Congress' Climate Apathy."
27. Kahn, "Hurricane Sandy Hasn't Shifted Climate Narrative." See also Marshall, *Don't Even Think About It*, chapter 2.
28. Cohen, *States of Denial*, p. 294.

## 7. The Five Psychological Barriers to Climate Action

1. Reviews are given by Stoknes, "Rethinking Climate Communications and the 'Psychological Climate Paradox'"; van der Linden, "Exploring Beliefs About Bottled Water and Intentions to Reduce Consumption."

## 8. From Barriers to Solutions

1. Not just that the Sámi have outboard engine and radios, of course, or that I had my advanced-materials backpack hanging from my shoulders. It was also that the air in

which the owls flew, even there, has been altered by the modernity. Their habitat is disturbed, too, though more subtly—by a degree or two Celsius (two to four degrees Fahrenheit). And finally, we were heading back toward home the next day, scheduled to reach "civilization" (meaning cars and buildings and roads) again the next night.

2. Hillman, *We've Had a Hundred Years of Psychotherapy—and the World's Getting Worse.*
3. Although no support was found for positive behavioral spillover, the study by Poortinga, Whitmarsh, and Suffolk, "The Introduction of a Single-Use Carrier Bag Charge in Wales," did find changes in self-reported environmental identity that could produce positive spillover effects in the longer term. This topic is further discussed in chapter 9.
4. Bain et al., "Promoting Pro-Environmental Action in Climate Change Deniers"; Stern, "Psychology: Fear and Hope in Climate Messages."
5. I'm all in favor of public regulation, too. In order to move the laggards, the slow adaptors, the resistors, regulation is sometimes the only thing that works. But the main thrust in communication should be positive, especially when introducing new regulations.
6. Paul Goodman, quoted in Hardin, "The Tragedy of the Commons."
7. Akerlof and Shiller, *Animal Spirits.*
8. Hawken, *Blessed Unrest.*
9. CRED Center for Research on Environmental Decisions (CRED), *The Psychology of Climate Change Communication*; Moser and Dilling, *Creating a Climate for Change*; Futerra Sustainability Communications, *Sell the Sizzle*; Kahan, "Making Climate-Science Communication Evidence-Based—All the Way Down"; Marshall, *Don't Even Think About It*, chapter 42.

## 9. The Power of Social Networks

1. Goldstein, "A Room with a Viewpoint." The new sign said: "Join your fellow guests in helping to save the environment. In a study conducted in fall 2003, 75% of the guests who stayed in this room . . . participated in our new resource savings program by using their towels more than once. You can join your fellow guests in this program to help save the environment by reusing your towels during your stay."
2. Griskevicius, Cantú, and Vugt, "The Evolutionary Bases for Sustainable Behavior"; Nolan et al., "Normative Social Influence Is Underdetected."
3. Manning, "The Effects of Subjective Norms on Behaviour in the Theory of Planned Behaviour."
4. Cialdini and Goldstein, "Social Influence"; Cialdini, Reno, and Kallgren, "A Focus Theory of Normative Conduct"; Cialdini and Rhoads, *Human Behavior and the Marketplace*; Cialdini, *Influence.*
5. Cialdini, Reno, and Kallgren, "A Focus Theory of Normative Conduct."
6. Rosenberg, *Join the Club*, p. 32.
7. Schultz, "Changing Behavior with Normative Feedback Interventions."
8. Cialdini, "Crafting Normative Messages to Protect the Environment"; Griskevicius, Cantú, and Vugt, "The Evolutionary Bases for Sustainable Behavior," p. 122.

9.  Ferraro and Price, "Using Non-Pecuniary Strategies to Influence Behavior."

10. Allcott, "Social Norms and Energy Conservation"; Ayres, Raseman, and Shih, "Evidence from Two Large Field Experiments That Peer Comparison Feedback Can Reduce Residential Energy Usage."

11. Tinjum, "Surging Forward."

12. See http://www.ecf.com/ news/b-track-b-the-best-marketing-tool-for-stimulating-cycling-cities.

13. Skogstad, "Tesla Model S." In the first half of 2014, three of the six bestselling cars were electric. See http://www.vg.no/forbruker/bil-baat-og-motor/bil-og-trafikk /hvilken-av-disse-tror-du-er-norges-mest-solgte-bil/a/23241389.

14. In fact, Opower's data suggests that on a percentage basis, areas with a low energy cost burden demonstrate even higher energy savings. Hallet and O'Brien, "It's Not All About the Money."

15. Sunstein, *Risk and Reason.*

16. Marshall, *How to Engage Your Community and Communicate About Climate Change.*

17. Ding et al., "Support for Climate Policy and Societal Action Are Linked to Perceptions About Scientific Agreement"; Maibach, Myers, and Leiserowitz, "Climate Scientists Need to Set the Record Straight."

18. Union of Concerned Scientists, *Cooler Smarter.*

19. From Adam Corner, http://talkingclimate.org/guides/using-social-norms -social-networks-to-promote-sustainable-behaviour. See also the analysis of the 2012 Obama reelection and marriage equality campaigns in Krygsman and ecoAmerica, *Campaigns II.*

20. Rosenberg, *Join the Club*, p. 285.

21. Krause and Basile, "Can Millennials and Social Networking Lead Us to a Sustainable Future?"

22. www.rinkwatch.org.

23. Paris, "Outdoor Rink Climate Change Project Gets Hundreds of Citizen Scientists."

24. Chris Dierker, market communications manager for Xcel Energy, presented preliminary survey data at ACEEE's Energy Efficiency as a Resource Conference in 2011. Vigen and Mazur-Stommen, "Reaching the 'High-Hanging Fruit' Through Behavior Change." See also Hamilton "Keeping Up with the Joneses" for research on UK experience with social learning on eco-renovation.

25. Bell and Weber, "We're Leaving Too Many Energy Dollars Behind Us, on the Ground"; Pash, "Coalition Brings Solar Energy Co-ops to Baltimore Area." See also American Solar Energy Society (ASES), "Cheaper by the Dozen."

26. See www.carbonconversations.org and Randall, "Loss and Climate Change."

27. Randall, chapter 17 in Rust and Totten, *Vital Signs.*

28. See McLean, "Buckinghamshire County Council Rolls Out Carbon Conversations"; Buchs, "'It Helped Me Sort of Face the End of the World.'"

29. For LEED data, see http://www.greensburgks.org and EarthGauge's report, http://www.earthgauge.net/wp-content/uploads/2009/03/eg_greensburg.pdf, accessed October 20, 2014.

30. Tony Leiserowitz, personal communication, September 2014.

31. Olli, Grendstad, and Wollebaek, "Correlates of Environmental Behaviors"; Nye and Burgess, *Promoting Durable Change in Household Waste and Energy Use Behaviour*.
32. Corner, "Social Norms & Social Networks."
33. Hawken, *Blessed Unrest*, p. 2.
34. Ibid., p. 4.
35. Jasanoff, "A New Climate for Society," p. 239.
36. Kahan, "Making Climate-Science Communication Evidence-Based," p. 15.
37. See Wolf and Moser, "Individual Understandings, Perceptions, and Engagement with Climate Change." They estimate direct energy use by US households at 38 percent of national emissions, more than any other country except China.
38. Truelove et al., "Positive and Negative Spillover of Pro-Environmental Behavior."
39. Phillips, "New Texas GOP Platform Calls on Politicians to Ignore Climate Change."

## 10. Reframing the Climate Messages

1. http://www.smh.com.au/news/environment/experts-warn-of-climate-mayhem /2007/03/29/1174761669481.html.
2. http://www.reuters.com/article/2014/02/16/ us-kerry-climate-idUSBREA1F0BP20140216.
3. http://accf.org/heated-rhetoric-doesnt-serve-debate-over-climate-change.
4. http://judithcurry.com/2013/05/31/rep-lamar-smith-on-climate-change.
5. http://www.washingtontimes.com/news/2014/feb/16/ politicians-agree-extreme-weather-costly-disagree-/#ixzz2wPC095UA.
6. http://www.telegraph.co.uk/earth/earthnews/3521567/People-must-be-willing-to -make-sacrifices-to-cut-climate-change.html.
7. Chapter 3 above introduced "framing" as referring to the implicit cognitive and semantic contexts around our concepts and conversations. Frames help construct meaning and understanding around a sentence or a message by employing the linguistic power of the metaphor. See Lakoff, *Don't Think of an Elephant!*
8. For all figures, see Painter, *Climate Change in the Media*, p. viii.
9. See Hart and Nisbet, "Boomerang Effects in Science Communication"; Myers et al., "A Public Health Frame Arouses Hopeful Emotions About Climate Change"; Nisbet, "Communicating Climate Change"; O'Neill and Nicholson-Cole, "'Fear Won't Do It.'"
10. DoD, Quadrennial Defense Review, 2014: "The pressures caused by climate change will influence resource competition while placing additional burdens on economies, societies, and governance institutions around the world. These effects are threat multipliers that will aggravate stressors abroad such as poverty, environmental degradation, political instability, and social tensions—conditions that can enable terrorist activity and other forms of violence," p. 8.
11. Painter, *Climate Change in the Media*, pp. 4–5.
12. For economic analyses of this critical question, I recommend the research of Harvard professor Martin Weitzman. In "On Modeling and Interpreting the Economics of Catastrophic Climate Change," p. 18, he says, "A crude natural metric for calibrating cost estimates of climate-change environmental-insurance policies might be that the

US already spends approximately 3 percent of national income on the cost of a clean environment." See also Weitzman, "GHG Targets as Insurance Against Catastrophic Climate Damages"; Weitzman "Fat-Tailed Uncertainty in the Economics of Catastrophic Climate Change."

13. Hartcher, "Cool Heads Missing in the Pressure Cooker." UNEP's (2011) Green Economy Report called for 2 percent of GDP investment per year to realize the green economy. This could be framed as an insurance policy against this climate risk.

14. Painter, *Climate Change in the Media*, p. 28.

15. Ibid., p. 27.

16. Bloomberg, Paulson, and Steyer, *Risky Business*.

17. Ibid.

18. Paulson, "The Coming Climate Crash."

19. Hurlstone et al., "Curbing Emissions."

20. Maibach et al., "Reframing Climate Change as a Public Health Issue"; Myers et al., "A Public Health Frame Arouses Hopeful Emotions About Climate Change."

21. Hopey, "Climate Change to Boost Health Problems."

22. Hamblin, "If You Have Allergies or Asthma, Talk to Your Doctor About Cap and Trade."

23. Patz et al., "Climate Change Challenges and Opportunities for Global Health."

24. Center for Research on Environmental Decisions (CRED), *The Psychology of Climate Change Communication*, p. 13.

25. Myers et al., "A Public Health Frame Arouses Hopeful Emotions About Climate Change," pp. 1110–11.

26. ecoAmerica, *Climate Impacts*; ecoAmerica, *New Facts, Old Myths*.

27. American Psychological Association's Task Force on the Interface Between Psychology and Global Climate Change, "Psychology and Global Climate Change," p. 115.

28. Crompton, *Common Cause*, chapter 2.

29. See Greaker, Stoknes, et al., "A Kantian Approach to Sustainable Development Indicators for Climate Change."

30. Global Commission on the Economy and Climate, *New Climate Economy Report*; Lovins, *Reinventing Fire*.

31. Bain et al., "Promoting Pro-Environmental Action in Climate Change Deniers."

32. Nisbet, "Communicating Climate Change," p. 20.

33. Kroh, "Push to Impose Extra Fees on Solar Customers Draws Outrage in Wisconsin."

34. Dembicki, "Global Shift to Clean Energy No Longer 'Theoretical.'"

35. Nuccitelli, "Can a Carbon Tax Work Without Hurting the Economy?"

36. Nuccitelli, "Let's Be Honest—The Global Warming Debate Isn't About Science."

37. UK Energy Research Centre, "Low Carbon Jobs"; SustainableBusiness.com, "US Solar Industry Employs More than Coal, Gas Industries Combined."

38. Wei, Patadia, and Kammen, "Putting Renewables and Energy Efficiency to Work."

39. The figures I quote here are in the self-interest of USGBC, of course, but for my purpose they illustrate the use of job-opportunity framings to replace doom framings. US Green Building Council (USGBC), "About USGBC."

40. Lakoff, "Why It Matters How We Frame the Environment," p. 79.
41. Hawken, Lovins, and Lovins, *Natural Capitalism*, 2010; Heck, Rogers, and Carroll, *Resource Revolution*; Weizsäcker, *Factor Five*.
42. Lakoff, "Why It Matters How We Frame the Environment"; Lakoff, *The All New Don't Think of an Elephant!*

## 11. Make It Simple to Choose Right

1. Johnson and Goldstein, "Do Defaults Save Lives?"; Pichert and Katsikopoulos, "Green Defaults."
2. Chetty et al., *Active vs. Passive Decisions and Crowdout in Retirement Savings Accounts*; Sunstein, "Behavioral Economics, Consumption, and Environmental Protection"; Thaler, *Nudge*.
3. Lorenzoni, Nicholson-Cole, and Whitmarsh, "Barriers Perceived to Engaging with Climate Change Among the UK Public and Their Policy Implications."
4. Starting in the summer of 2007. http://www.nbcs.rutgers.edu/ccf/main/print.
5. Egebark and Ekström, *Can Indifference Make the World Greener?*
6. It would be more effective as a compulsory tax, of course, but assuming that tax is not politically feasible, then attempting a default nudge doesn't require lengthy political decisions.
7. Because of restriction on private cars in the inner city during rush hour, the Bogotá buses are running three times as fast as a typical New York bus. In Bogotá vehicle traffic is reduced by 40 percent through a measure that restricts vehicles, according to the last number in their license plate, from traveling during peak hours in the entire urban area. In addition, car-free Sundays and the removal of thousands of parking spots have made the roads more pedestrian-friendly. On car-free Sundays the city is now using public streets as a large open park. Bogotá has become a healthier and safer city with greater social integration and cheap, sustainable transport. http://www.dac.dk/en/dac-cities/sustainable-cities/all-cases/transport/bogota-more-bikes-and-buses-fewer-cars.
8. Kallbekken, Sælen, and Hermansen, "Bridging the Energy Efficiency Gap."
9. Ibid.
10. Bennhold, "Britain's Ministry of Nudges."
11. Pichert and Katsikopoulos, "Green Defaults."
12. Cho, "Making Green Behavior Automatic."
13. Kallbekken and Sælen, "'Nudging' Hotel Guests to Reduce Food Waste as a Win–Win Environmental Measure."
14. Cho, "Making Green Behavior Automatic."
15. Hohle, "'Nudging' Sustainable Food Choices."
16. See the Vest Pizza campaign video: https://www.youtube.com/watch?v=dxNaWKrIRo0.
17. Fischer, "Clever Campaign in Denmark: If You Save Energy, You Get Free Pizza." According to strict definitions of the word *nudge*, this may not be one: It offers tangible rewards for actions and thus is an incentive scheme. Still, I've included it because it does attempt behavioral modification through interactive informational design.

18. http://buildingsdatabook.eren.doe.gov.
19. http://www.tu.no/kraft/2014/06/03/da-kundene-matte-betale-for-effekt-i-stedet
    -for-forbruk-gikk-stromforbruket-ned-med-20-prosent.
20. Johnson et al., "Beyond Nudges"; Sunstein, "Behavioral Economics, Consumption,
    and Environmental Protection"; Thaler, *Nudge*.
21. Treuer et al., "Weathering the Storm."
22. Ibid; Aas, Minken, and Samstad. *Myter og fakta om køprising*.

## 12. Use the Power of Stories to Re-Story Climate

1. Hulme, *Why We Disagree About Climate Change*; Painter, *Climate Change in the
   Media*; Stoknes, *Sjelens Landskap*. An English translation of the chapter on the
   psychology of apocalypse is available at http://www.wildethics.com/essays
   /relating-to-climate-change.html, accessed October 1, 2014.
2. O'Neill and Nicholson-Cole, "'Fear Won't Do It.'"
3. http://www.dailykos.com/story/2014/05/15/1299514/-Stephen-Colbert-on-our
   -failure-to-take-action-on-climate-change#.
4. Corner, "The 'Art' of Climate Change Communication."
5. Monbiot, *Feral*; Harré, *Psychology for a Better World*.
6. McDonough, *The Upcycle*; Pauli, *The Blue Economy*; Porritt, *The World We Made*;
   Sustainia, "Sustainable Solutions"; World Business Council for Sustainable
   Development (WBCSD), *Vision 2050*.
7. Randall, "Loss and Climate Change"; Stoknes, "Rethinking Climate Communications
   and the 'Psychological Climate Paradox.'"
8. Espinasse, Francis. *Lancashire Worthies*. London: Simpkin, Marshall, 1874.
9. *Economist*, "Commodity Prices in the (Very) Long Run"; Jacks, "From Boom to
   Bust"; McKinsey Global Institute, "Resource Revolution."
10. Rockström et al., "A Safe Operating Space for Humanity."
11. Heck, Rogers, and Carroll, *Resource Revolution*; Lovins, *Reinventing Fire*, chapter 2.
12. Source: United Nations Environment Program (UNEP), http://www.unep.org
    /wed/2013/docs/Waste_Less_Lunch_presentation.pdf. See also Food and Agriculture
    Organization (FAO), *Food Wastage Footprint*.
13. United Nations Environment Program (UNEP), Regional Office of North America,
    "Food Waste: The Facts," http://www.worldfooddayusa.org/food_waste_the_facts.
14. Weizsäcker, *Factor Five*, chapter 3, p. 155.
15. Lovins, *Reinventing Fire*.
16. See Global e-Sustainability Initiative (GeSI), *GeSI SMARTer2020*, p. 12.
17. Greenpeace, *Clicking Clean*.
18. The infamous Jevons paradox and rebound effect are not destiny. If technology
    shifts are accompanied by institutional innovations at the same time, the rebound in
    throughput of materials can be restrained. See Weizsäcker, *Factor Five*, chapter 8.
19. Of course, incumbents and some old resource owners would want—from narrow
    self-interest—to keep the waste and thus prices up. But they can't publicly say so.
20. Futerra Sustainability Communications, *Sell the Sizzle*, p. 10.

21. Lucas, *Lectures on Economic Growth.*
22. Piketty, *Capital in the Twenty-First Century.*
23. In the United States, median income has not grown over the last decades, which may explain parts of this conundrum. See Stiglitz, Sen, and Fitoussi, *Report of the Commission on the Measurement of Economic Performance and Social Progress (CMEPSP).*
24. Keynes, "Economic Possibilities for Our Grandchildren."
25. White, "The Historical Roots of Our Ecologic Crisis."
26. Palmer, *Faith in Conservation.*
27. Brown, "The Rise of Ecothology."
28. Casey, "Straight Talk About Radical Climate Change—and How to Stop It."
29. Jenkins, "Pope Francis Makes Biblical Case for Addressing Climate Change."
30. Leber, "76 Percent of Religious Americans Want a Global Pact Cutting Pollution, Viewing It in Moral and Religious Terms."
31. Gardiner and Hartzell-Nichols, "Ethics and Global Climate Change."
32. Corner, *A New Conversation with the Centre-Right About Climate Change.*
33. The proposed amendment to the Rome Statute, by Polly Higgins, April 2010, from http://eradicatingecocide.com/overview/what-is-ecocide/.
34. Monbiot, *Feral*, p. 12.
35. Groves, "Mountain Lion Makes Itself at Home in Griffith Park."
36. Gibson, *A Reenchanted World.*
37. Stephenson et al., "Rate of Tree Carbon Accumulation Increases Continuously with Tree Size."
38. For sources on climate and personal storytelling, see: Randall, Salsbury, and White, *Moving Stories*; Corner and van Eck, *Science & Stories*; Center for Research on Environmental Decisions (CRED), *The Psychology of Climate Change Communication*; Harré, *Psychology for a Better World*, pp. 52–56; Marshall, *Don't Even Think About It*, chapter 20; Moser and Dilling, *Creating a Climate for Change*; Olson, *Don't Be Such a Scientist*; The Story Group, *National Climate Assessment Videos.*
39. Sunstein, "Behavioral Economics, Consumption, and Environmental Protection."
40. Painter, *Climate Change in the Media*, p. 8.

## 13. New Signals of Progress

1. Joseph Stiglitz, "Progress, What Progress?"
2. United Nations University and United Nations Environment Program (UNEP), *Inclusive Wealth Report 2012.*
3. Hulme, *Why We Disagree About Climate Change*, p. 334.
4. The company Opower is providing individual feedback on exactly this; see chapter 7. See also Rosentrater et al., "Efficacy Trade-Offs in Individuals' Support for Climate Change Policies."
5. For two examples of such indicator overviews, see Organization for Economic Cooperation and Development (OECD), *Green Growth Indicators 2014*; Stiglitz, Sen, and Fitoussi, *Report of the Commission on the Measurement of Economic Performance and Social Progress (CMEPSP).*

6. Randers, "Greenhouse Gas Emissions per Unit of Value Added ('GEVA')—A Corporate Guide to Voluntary Climate Action."

7. Alliance to Save Energy, "Energy 2030"; Hawken, Lovins, and Lovins, *Natural Capitalism*, 2005; McKinsey & Company, *Impact of the Financial Crisis on Carbon Economics*; WWF, CDP, and McKinsey & Company, *The 3% Solution*.

8. Organization for Economic Cooperation and Development (OECD), *Green Growth Indicators 2014*, p. 58.

9. Stiglitz, Sen, and Fitoussi, *Report of the Commission on the Measurement of Economic Performance and Social Progress (CMEPSP)*; Kahneman et al., *Well-Being*.

10. Ibid.

11. Dasgupta is science adviser and lead author of: United Nations University and United Nations Environment Program (UNEP), *Inclusive Wealth Report 2012*.

12. "Act only according to that maxim whereby you can at the same time will that it should become a universal law without contradiction." Kant, *Grounding for the Metaphysics of Morals*, p. 30.

13. Around 4.5 tons per year per person is $CO_2$ from fossil fuels, while another 2.5 tons is all other anthropogenic greenhouse gas emissions, mainly land use in forestry and agriculture. See Intergovernmental Panel on Climate Change (IPCC) Working Group 3, *Climate Change 2013: Mitigation of Climate Change: Summary for Policymakers*, figure SPM 1.

14. Greaker et al., "A Kantian Approach to Sustainable Development Indicators for Climate Change."

15. "Contrary to the UK government's conclusions, most research shows personal carbon trading to be at least as socially acceptable as an alternative taxation policy. People think it could be both fair and effective." Fawcett, "Personal Carbon Trading." See also http://www.teqs.net for a broader introduction to this. In a "cap and dividend" system, public revenue raised from the sale of pollution credits is rebated to citizens or to consumers.

16. *The Millennium Ecosystem Assessment Report, 2005*, is now being updated by the Intergovernmental Platform on Biodiversity and Ecosystem Services (IPBES)—an independent body modeled on the Intergovernmental Panel on Climate Change.

## 14. The Air's Way of Being

1. Abram, "Climate and Psyche."

2. Shaw, *A Branch from the Lightning Tree*, p. 70.

3. This has been mainstream economists' line of reasoning, as for instance in Nordhaus, *A Question of Balance*.

4. For a psychological perspective, see the essay "The Imagination of Air" from 1981 in Hillman, *Alchemical Psychology*, chapter 9. For a historical perspective, see McEvoy, *The Historiography of the Chemical Revolution*.

5. It is important to be aware that the air consists of gases that react chemically with one another, and is at a so-called constant chemical dis-equilibrium: For instance, methane and oxygen would soon react with each other and change the composition

if they both weren't replenished continually. Who does the replenishment? Algae and plants replenish the oxygen via photosynthesis, and keep it in reactive state that is suitable for life. This is the main idea in Gaia theory by James Lovelock. See Lovelock, *Gaia*; Lovelock and Margulis, "Atmospheric Homeostasis by and for the Biosphere."

6. See, for instance, Csank et al., "Estimates of Arctic Land Surface Temperatures During the Early Pliocene from Two Novel Proxies."
7. *New Scientist*, "Climate Change: Awaking the Sleeping Giants."
8. Wystan Hugh Auden, "In Memory of Ernst Toller," originally published in *Another Time* (1940).

## 15. Stand Up for Your Depression!

1. *International New York Times* Editorial Board, "The Disappearing Moose"; Wilcove and Curwood, "Messed Up Migrations."
2. This is one of my friend Stephan Harding's favorite tongue-in-cheek punch lines when teaching Gaia theory and earth system science.
3. Lertzman, "The Myth of Apathy," p. 130.
4. Bly, Hillman, and Meade, *The Rag and Bone Shop of the Heart*, 96.
5. Harbert, "In Despair Over the Polar Bear."
6. In Norgaard and Reed, "Emotional Impacts of Environmental Decline."
7. Wildcat, *Red Alert!*, chapter 1.
8. Albrecht et al., "Solastalgia"; Smith, "Is There an Ecological Unconscious?"
9. Wilson, *Coal Blooded*.
10. Clayton, Manning, and Hodge, *Beyond Storms & Droughts*; Doherty and Clayton, "The Psychological Impacts of Global Climate Change"; Weissbecker, *Climate Change and Human Well-Being*.
11. Lifton, "Beyond Psychic Numbing."
12. Hillman, *The Thought of the Heart and The Soul of the World*.
13. Berry, *A Timbered Choir*, p. 98.
14. Phyllis Windle, "The Ecology of Grief," in Roszak, Gomes, and Kanner, *Ecopsychology*.
15. Joanna Macy, "Working Through Environmental Despair," in ibid.
16. Joanna Macy in Rust and Totten, *Vital Signs*.
17. Hillman, *We've Had a Hundred Years of Psychotherapy—and the World's Getting Worse*, p. 98. As a former clinical psychologist, I'd also point out that there are severe types of clinical depressions that cripple the patient. Neither Hillman nor I are speaking of this most severe type here.
18. Ibid., p. 45, italics mine.
19. Crutzen, "Geology of Mankind."
20. Haberl, Erb, and Krausmann, "Global Human Appropriation of Net Primary Production (HANPP)"; Krausmann et al., "Global Human Appropriation of Net Primary Production Doubled in the 20th Century."
21. The Dark Mountain Project has a website, and its manifesto can be read here: http://dark-mountain.net/about/manifesto.

22. Greenpeace, "Help Save Santa's Home This Christmas."
23. See for instance Gibbons, "Time to Sign up for the Climate Change War."
24. John Steiner, "Discussion—Climate Change in a Perverse Culture," p. 80 in Weintrobe, *Engaging with Climate Change*.
25. Ibid.
26. Bernstein, *Living in the Borderland*.
27. Leopold, *A Sand County Almanac*.
28. Jensen, "Beyond Hope."
29. Hopkins, "Paul Kingsnorth on Living with Climate Change."
30. The Kübler-Ross stages are often dubbed DABDA: denial, anger, bargaining, depression, acceptance. Kübler-Ross and Kessler, *On Grief and Grieving*. A newer, modified grief model can found in Worden, *Grief Counseling and Grief Therapy*.
31. Randall, "Loss and Climate Change"; Running, "Five Stages of Climate Grief."
32. Wysham, "The Six Stages of Climate Grief."
33. See a sociological analysis of this in Gibson, *A Reenchanted World*.
34. Noel, *The Soul of Shamanism*.
35. Wilson, *Biophilia*.
36. Haukeland, *Dyp Glede*.
37. Jensen, "Beyond Hope."
38. Hillman, *Re-Visioning Psychology*, part 4.

## 16. Climate Disruption as Symptom: What Is It Trying to Tell Us?

1. Moore, *Care of the Soul*.
2. Freud, *Civilization and Its Discontents*.
3. From "Anima Mundi" in Hillman, *City & Soul*, p. 31.
4. Hillman, *Re-Visioning Psychology*, p. 55.
5. Hillman, *Re-Visioning Psychology*, p. 82ff.
6. Fisher, *Radical Ecopsychology*; Hillman, *City & Soul*; Kahn and Hasbach, *Ecopsychology*; Roszak, Gomes, and Kanner, *Ecopsychology*; Rust and Totten, *Vital Signs*.
7. The classic reference is Argyris, *Organizational Learning*.
8. Hawken, Lovins, and Lovins, *Natural Capitalism*, 1999; McDonough, *The Upcycle*; Weizsäcker, *Factor Five*.
9. Næss, *Ecology, Community, and Lifestyle*.
10. Dreyfus, *Being-in-the-World*.
11. Clive Hamilton in Weintrobe, *Engaging with Climate Change*.
12. I'll address this dualistic mode of thinking below in the section on Understanding the Limitations of Science and Psychology, as well as in the next chapter.
13. Fisher, *Radical Ecopsychology*, chapter 2; von Wright, *Explanation and Understanding*.
14. Randers, *2052*; Rockström et al., "A Safe Operating Space for Humanity"; Wackernagel, "Global Footprint Network: Ecological Footprint."
15. Dunlap, "The New Environmental Paradigm Scale"; Hernes, *Hot Topic—Cold Comfort*; Latour, *We Have Never Been Modern*; Latour, "Waiting for Gaia."

16. Latour, *We Have Never Been Modern*.
17. Sometimes a capital *S* is used for the larger, ecological Self to distinguish it from the individual, smaller self. Clayton and Opotow, *Identity and the Natural Environment*, pp. 45–46; Fisher, *Radical Ecopsychology*; Roszak, Gomes, and Kanner, *Ecopsychology*; Stoknes, "Økopsykologi: Fra et intrapsykisk til et økologisk Selv."
18. Quoted in Fisher, *Radical Ecopsychology*, p. 94.
19. King, "A Critique of Western Psychology from an American Indian Psychologist."
20. Mayer and Frantz, "The Connectedness to Nature Scale"; Schultz et al., "Implicit Connections with Nature"; Tam, "Concepts and Measures Related to Connection to Nature"; Stern, "New Environmental Theories."
21. Crompton and Kasser, *Meeting Environmental Challenges*; Schultz et al., "Implicit Connections with Nature," p. 32; Stern, "New Environmental Theories."
22. Crompton and Kasser, *Meeting Environmental Challenges*, p. 9.
23. Stoknes, *Money & Soul*.
24. Latour, *We Have Never Been Modern*.
25. "From Mirror to Window" in Hillman, *City & Soul*, p. 66.
26. "Anima Mundi: Return of the Soul to the World" in ibid., p. 27.
27. Box, "Is the Climate Dragon Awakening?"
28. *New Scientist*, "Climate Change: Awaking the Sleeping Giants."

## 17. Re-Imagining Climate as the Living Air

1. A term first improvised by Martin Heidegger in *Being and Time*, 1927.
2. Hillman in *Re-Visioning Psychology*, p. 15.
3. Harrison, *Elements of Pantheism*, 78.
4. *New Oxford American Dictionary*.
5. Jung, *The Collected Works of C. G. Jung*, p. 51, CW13.
6. Ibid., CW6, §78.
7. Abram, *The Spell of the Sensuous*.
8. Merleau-Ponty, *Phenomenology of Perception*, p. xi.
9. Abram, *Becoming Animal*, p. 136.
10. Van der Post, *The Heart of the Hunter*, p. 91.
11. Brooke, *Jung and Phenomenology*, p. 85.
12. Heidegger, *Basic Writings*, pp. 149–87.
13. Genesis 2:7 (King James Version).
14. Archetypal psychologist Tom Cheetham relates what Sufi scholar Henry Corbin says about Suzuki, the Zen Buddhist, interacting with a spoon: "This spoon now exists in Paradise . . . We are now in Heaven." Cheetham, *Green Man, Earth Angel*, p. 11.
15. Matthew 4:17, God's Word translation.
16. McNeley, *Holy Wind in Navajo Philosophy*.
17. Ibid., pp. 50–51.
18. Ibid., pp. 30–31.
19. Abram, *The Spell of the Sensuous*, p. 237.
20. Ibid.

21. With *image* I refer to any patterning of information—visual, auditory, olfactory, or ideational—not just to visual images.

## 18. It's Hopeless and I'll Give It My All

1. http://theconversation.com/we-can-now-only-watch-as-west-antarcticas-ice-sheets-collapse-26957 and http://www.reuters.com/article/2014/05/12/us-climatechange-antarctica-idUSKBN0DS1IH20140512.
2. Shaw, *A Branch from the Lightning Tree.*
3. Hillman, *The Soul's Code.*
4. Joanna Macy also distinguishes between passive and active varieties of hope in her book *Active Hope.*
5. Within later systems theory, the term *autopoietic* is often used for self-healing or self-caring. Rust and Totten, *Vital Signs*, p. 44. We're part of the earth's autopoesis.
6. Hamilton, "What History Can Teach Us About Climate Change Denial," p. 28.
7. Camus, *The Plague*, p. 255.

# —BIBLIOGRAPHY—

Aakre, S., and J. Hovi. "Emission Trading: Participation Enforcement Determines the Need for Compliance Enforcement." *European Union Politics* 11, no. 3 (September 1, 2010): 427–45. doi:10.1177/1465116510369265.

Aas, Harald, Harald Minken, and Hanne Samstad. *Myter og fakta om køprising* (Institute of Transport Economics [TØI] of the Norwegian Centre for Transport Research Report 1010/2009). Oslo: TØI, March 2009. https://www.toi.no/getfile.php/ Publikasjoner/T%C3%98I%20rapporter/2009/1010-2009/1010-2009-nett.pdf.

Abram, David. *Becoming Animal: An Earthly Cosmology.* New York: Vintage Books, 2011.

———. "Climate and Psyche." Speech presented at the Climate Psychology seminar, Norwegian Business School, Oslo, September 28, 2012. http://www.youtube.com/ watch?v=TzZ41o3VU7M.

———. *The Spell of the Sensuous: Perception and Language in a More-than-Human World.* New York: Vintage Books, 1997.

Akerlof, George A., and Robert J. Shiller. *Animal Spirits: How Human Psychology Drives the Economy, and Why It Matters for Global Capitalism.* Princeton, NJ: Princeton University Press, 2009.

Albrecht, Glenn, Gina-Maree Sartore, Linda Connor, Nick Higginbotham, Sonia Freeman, Brian Kelly, Helen Stain, Anne Tonna, and Georgia Pollard. "Solastalgia: The Distress Caused by Environmental Change." *Australasian Psychiatry* 15, no. s1 (January 2007): S95–98. doi:10.1080/10398560701701288.

Allcott, Hunt. "Social Norms and Energy Conservation." *Journal of Public Economics* 95, no. 9–10 (October 2011): 1082–95. doi:10.1016/j.jpubeco.2011.03.003.

Alliance to Save Energy (ASE). "Energy 2030 Goal." ASE (website), March 17, 2014. http:// www.ase.org/policy/energy2030.

American Psychological Association's Task Force on the Interface Between Psychology and Global Climate Change. "Psychology and Global Climate Change: Addressing a Multi-Faceted Phenomenon and Set of Challenges." APA, 2009. http://www.apa.org/ science/about/publications/climate-change.aspx.

American Solar Energy Society (ASES). "Cheaper by the Dozen: Solar Group Purchase Programs." *Solar Today*, May 9, 2013. http://www.ases.org/2013/05/ cheaper-by-the-dozen-solar-group-purchase-programs.

Amos, Jonathan. "Arctic Summers Ice-Free 'by 2013.'" *BBC News*, December 12, 2007. http://news.bbc.co.uk/2/hi/7139797.stm.

Anderegg, W. R. L., J. W. Prall, J. Harold, and S. H. Schneider. "Expert Credibility in Climate Change." *Proceedings of the National Academy of Sciences* 107, no. 27 (June 21, 2010): 12107–09. doi:10.1073/pnas.1003187107.

Argyris, Chris. *Organizational Learning.* Addison-Wesley OD Series. Reading, MA: Addison-Wesley, 1978.

Ariely, Dan. *Predictably Irrational: The Hidden Forces That Shape Our Decisions*. 1st ed. New York: Harper, 2008.

Aristotle. *Aristotle's Nicomachean Ethics*. Chicago and London: University of Chicago Press, 2011.

Austgulen, Marthe. *Nordmenns Holdninger Til Klimaendringer, Medier Og Politikk*. Project note. Oslo: Staten Institutt for Forbruksforskning (SIFO), 2012. http://www.sifo.no/files/file78208_rapport_climate_crossroads_web.pdf.

Austgulen, Marthe, and Eivind Stø. "Klimaskepsis I Norge." Oslo: Statens Institutt for Forbruksforskning (SIFO), 2012. http://www.sifo.no/files/file78113_presentasjon_offentlig_9-mai.pdf.

Ayres, I., S. Raseman, and A. Shih. "Evidence from Two Large Field Experiments That Peer Comparison Feedback Can Reduce Residential Energy Usage." *Journal of Law, Economics, and Organization* 29, no. 5 (August 20, 2012): 992–1022. doi:10.1093/jleo/ews020.

Bagley, Katherine. "Q&A: How a SuperPAC on a Shoestring Is Taking on Congress' Climate Apathy." *InsideClimate News*, August 20, 2014. http://insideclimatenews.org/news/20140820/qa-how-superpac-shoestring-taking-congress-climate-apathy.

Bain, Paul G., Matthew J. Hornsey, Renata Bongiorno, and Carla Jeffries. "Promoting Pro-Environmental Action in Climate Change Deniers." *Nature Climate Change* 2, no. 8 (June 17, 2012): 600–03. doi:10.1038/nclimate1532.

Ball, Tim F. *The Deliberate Corruption of Climate Science*. Mount Vernon, WA: Stairway Press, 2014.

Barclay, Pat. "Trustworthiness and Competitive Altruism Can Also Solve the 'Tragedy of the Commons.'" *Evolution and Human Behavior* 25, no. 4 (July 2004): 209–20. doi:10.1016/j.evolhumbehav.2004.04.002.

Bateson, Gregory, and Mary Catherine Bateson. *Angels Fear: Towards an Epistemology of the Sacred*. Toronto and New York: Bantam, 1988.

Bell, Larry. *Climate of Corruption: Politics and Power Behind the Global Warming Hoax*. Austin, TX: Greenleaf Book Group Press, 2011.

———. "Global Warming Alarm: Continued Cooling May Jeopardize Climate Science and Green Energy Funding!" *Forbes*, April 30, 2013. http://www.forbes.com/sites/larrybell/2013/04/30/global-warming-alarm-continued-cooling-may-jeopardize-climate-science-and-green-energy-funding.

———. "The Feverish Hunt for Evidence of a Man-Made Global Warming Crisis." *Forbes*, March 19, 2013. http://www.forbes.com/sites/larrybell/2013/03/19/the-feverish-hunt-for-evidence-of-a-man-made-global-warming-crisis.

Bell, Ruth Greenstad, and Elke U. Weber. "Opinion: We're Leaving Too Many Energy Dollars Behind Us, on the Ground." *Daily Climate*, May 19, 2014. http://www.dailyclimate.org/tdc-newsroom/2014/05/saving-energy.

Benestad, Rasmus. "A Failure in Communicating the Impact of New Findings." *RealClimate*, December 6, 2013. http://www.realclimate.org/index.php/archives/2013/12/a-failure-in-communicating-the-impact-of-new-findings.

Bennhold, Katrin. "Britain's Ministry of Nudges." *New York Times*, December 7, 2013. http://www.nytimes.com/2013/12/08/business/international/britains-ministry-of-nudges.html?pagewanted=4&pagewanted=all&_r=0.

Bensinger, Ken, and Ralph Vartabedian. "Toyota to Fix 'Very Dangerous' Gas Pedal Defects." *Los Angeles Times*, November 26, 2009. http://articles.latimes.com/2009/nov/26/business/la-fi-toyota-recall26-2009nov26.

Bernstein, Jerome S. *Living in the Borderland: The Evolution of Consciousness and the Challenge of Healing Trauma*. East Sussex, UK, and New York: Brunner-Routledge, 2005.

Berry, Wendell. *A Timbered Choir: The Sabbath Poems, 1979–1997*. Washington, DC: Counterpoint, 1998.

Bloomberg, Michael, Henry Paulson, and Thomas Steyer. Risky Business: The Economic Risk of Climate Change in the US (website of the Risky Business Project), June 2014. http://riskybusiness.org.

Bly, Robert, James Hillman, and Michael Meade. *The Rag and Bone Shop of the Heart: Poems for Men*. New York: HarperCollins, 1993.

Bojinski, Stephan, Michel Verstraete, Thomas C. Peterson, Carolin Richter, Adrian Simmons, and Michael Zemp. "The Concept of Essential Climate Variables in Support of Climate Research, Applications, and Policy." *Bulletin of the American Meteorological Society* 95, no. 9 (September 2014): 1431–43. doi:10.1175/BAMS-D-13-00047.1.

Box, Jason. "Is the Climate Dragon Awakening?" *Meltfactor* (blog), July 27, 2014. http://www.meltfactor.org/blog/?p=1329.

Britt, Robert R. "The Odds of Dying." *LiveScience*, January 6, 2005. http://www.livescience.com/3780-odds-dying.html.

Brooke, Roger. *Jung and Phenomenology*. Pittsburgh, PA: Trivium Publications, 2009.

Brown, Valerie. "The Rise of Ecothology." *21stC* 3, no. 4 (May 17, 2014). http://www.columbia.edu/cu/21stC/issue-3.4/brown.html.

Brulle, Robert J. "Institutionalizing Delay: Foundation Funding and the Creation of US Climate Change Counter-Movement Organizations." *Climatic Change*, December 21, 2013. doi:10.1007/s10584-013-1018-7.

Brulle, Robert J., Jason Carmichael, and J. Craig Jenkins. "Shifting Public Opinion on Climate Change: An Empirical Assessment of Factors Influencing Concern over Climate Change in the US, 2002–2010." *Climatic Change* 114, no. 2 (February 3, 2012): 169–88. doi:10.1007/s10584-012-0403-y. .

Brysse, Keynyn, Naomi Oreskes, Jessica O'Reilly, and Michael Oppenheimer. "Climate Change Prediction: Erring on the Side of Least Drama?" *Global Environmental Change* 23, no. 1 (February 2013): 327–37. doi:10.1016/j.gloenvcha.2012.10.008.

Buchs, Milena. "'It Helped Me Sort of Face the End of the World': The Role of Emotions for Third Sector Climate Change Engagement Initiatives." *Environmental Values*, in press (2014).

Buckels, Erin E., Paul D. Trapnell, and Delroy L. Paulhus. "Trolls Just Want to Have Fun." *Personality and Individual Differences* 67 (September 2014): 97–102. doi:10.1016/j.paid.2014.01.016.

Camus, Albert. *The Plague*. 1st Vintage international ed. New York: Vintage Books, 1991.

Casey, Tina. "Straight Talk About Radical Climate Change—and How to Stop It." *CleanTechnica*, November 18, 2013. http://cleantechnica.com/2013/11/18/kibbutz-ketura-israel-solar-power.

Center for Research on Environmental Decisions (CRED). *The Psychology of Climate Change Communication: A Guide for Scientists, Journalists, Educators, Aides and the Interested Public.* New York: CRED, 2009.

Cheetham, Tom. *Green Man, Earth Angel: The Prophetic Tradition and the Battle for the Soul of the World.* SUNY Series in Western Esoteric Traditions. Albany, NY: State University of New York Press, 2005.

Chetty, Raj, John N. Friedman, Soren Leth-Petersen, Torben Nielsen, and Tore Olsen. *Active vs. Passive Decisions and Crowdout in Retirement Savings Accounts: Evidence from Denmark* (NBER Working Paper No. 18565). National Bureau of Economic Research, 2012.

Cho, Renee. "Making Green Behavior Automatic." *State of the Planet* (blog of the Earth Institute), May 23, 2013. http://blogs.ei.columbia.edu/2013/05/23/making-green-behavior-automatic.

Cialdini, Robert B. "Crafting Normative Messages to Protect the Environment." *Current Directions in Psychological Science* 12, no. 4 (August 1, 2003): 105–09. doi:10.2307/20182853.

———. *Influence: The Psychology of Persuasion.* New York: Collins, 2007.

Cialdini, Robert B., and Noah J. Goldstein. "Social Influence: Compliance and Conformity." *Annual Review of Psychology* 55, no. 1 (February 2004): 591–621. doi:10.1146/annurev.psych.55.090902.142015.

Cialdini, Robert B., Raymond R. Reno, and Carl A. Kallgren. "A Focus Theory of Normative Conduct: Recycling the Concept of Norms to Reduce Littering in Public Places." *Journal of Personality and Social Psychology* 58, no. 6 (1990): 1015–26. doi:10.1037/0022-3514.58.6.1015.

Cialdini, Robert B., and Kelton Rhoads. *Human Behavior and the Marketplace.* American Marketing Association, 2001. http://www.influenceatwork.com/wp-content/uploads/2012/02/Marketing_Research.pdf.

Clayton, Susan, C. M. Manning, and Caroline Hodge. *Beyond Storms & Droughts: The Psychological Impacts of Climate Change.* Washington, DC: American Psychological Association and EcoAmerica, 2014.

Clayton, Susan D., and Susan Opotow, eds. *Identity and the Natural Environment: The Psychological Significance of Nature.* Cambridge, MA: MIT Press, 2003.

Cohen, Stanley. *States of Denial: Knowing About Atrocities and Suffering.* Cambridge, UK, and Malden, MA: Polity and Blackwell Publishers, 2001.

Committee for a Constructive Tomorrow (CFACT). "Release: Call to Suspend Climate Treaty Negotiations at CFACT Press Conference." CFACT, December 6, 2012. http://www.cfact.org/2012/12/06/call-to-suspend-climate-treaty-negotiations-at-cfact-press-release-more.

Cook, John. "Attacks on Scientific Consensus on Climate Change Mirror Tactics of Tobacco Industry." *Skeptical Science*, November 28, 2013. http://www.skepticalscience.com/Attacks-scientific-consensus-mirror-tobacco-industry.html.

Cook, John, Dana Nuccitelli, Sarah A. Green, Mark Richardson, Bärbel Winkler, Rob Painting, Robert Way, Peter Jacobs, and Andrew Skuce. "Quantifying the Consensus on Anthropogenic Global Warming in the Scientific Literature." *Environmental Research Letters* 8, no. 2 (June 1, 2013). doi:10.1088/1748-9326/8/2/024024.

Corner, Adam. "Social Norms & Social Networks." *Talking Climate*, 2012. http://talking climate.org/guides/using-social-norms-social-networks-to-promote-sustainable -behaviour.

———. *A New Conversation with the Centre-Right About Climate Change: Values, Frames and Narratives*. Climate Outreach Information Network (COIN), 2013. http://talkingclimate.org/wp-content/uploads/2013/06/COIN-A-new-conversation -with-the-centre-right-about-climate-change.pdf.

———. "The 'Art' of Climate Change Communication." *The Guardian*, March 18, 2013. http://www.theguardian.com/sustainable-business/art-climate-change -communication.

Corner, Adam, and Christel van Eck. *Science & Stories: Bringing the IPCC to Life*. Climate Outreach Information Network (COIN), May 2014. www.climateoutreach.org.uk.

Crompton, Tom. *Common Cause: The Case for Working with Our Cultural Values*. World Wildlife Fund UK, September 2010. http://assets.wwf.org.uk/downloads /common_cause_report.pdf.

Crompton, Tom, and Tim Kasser. *Meeting Environmental Challenges: The Role of Human Identity*. WWF-UK's Strategies for Change Project. World Wildlife Fund UK, 2009. http://assets.wwf.org.uk/downloads/meeting_environmental_challenges___the _role_of_human_identity.pdf.

Crutzen, Paul J. "Geology of Mankind." *Nature* 415, no. 6867 (January 3, 2002): 23. doi:10.1038/415023a.

Csank, Adam Z., Aradhna K. Tripati, William P. Patterson, Robert A. Eagle, Natalia Rybczynski, Ashley P. Ballantyne, and John M. Eiler. "Estimates of Arctic Land Surface Temperatures During the Early Pliocene from Two Novel Proxies." *Earth and Planetary Science Letters* 304, no. 3–4 (April 2011): 291–99. doi:10.1016/j. epsl.2011.02.030.

Curry, Judith. "Uncertainty in Sea Surface Temperature Measurements and Data Sets." *Climate Etc.* (blog), November 13, 2013. http://judithcurry.com/2013/11/13 /uncertainty-in-sst-measurements-and-data-sets.

Dawkins, Richard. *The Selfish Gene*. New ed. Oxford, UK, and New York: Oxford University Press, 1989.

Dembicki, Geoff. "Global Shift to Clean Energy No Longer 'Theoretical.'" *The Tyee*, February 22, 2014. http://thetyee.ca/News/2014/02/22/Clean-Energy-Global-Shift.

Diamond, Jared M. *Collapse: How Societies Choose to Fail or Succeed*. New York: Penguin Books, 2011.

Diethelm, P., and M. McKee. "Denialism: What Is It and How Should Scientists Respond?" *European Journal of Public Health* 19, no. 1 (October 16, 2008): 2–4. doi:10.1093 /eurpub/ckn139.

Ding, Ding, Edward W. Maibach, Xiaoquan Zhao, Connie Roser-Renouf, and Anthony Leiserowitz. "Support for Climate Policy and Societal Action Are Linked to Perceptions About Scientific Agreement." *Nature Climate Change* 1, no. 9 (November 20, 2011): 462–66. doi:10.1038/nclimate1295.

Doherty, Thomas J., and Susan Clayton. "The Psychological Impacts of Global Climate Change." *American Psychologist* 66, no. 4 (2011): 265–76. doi:10.1037/a0023141.

Donner, Simon. "Talking About Climate Change in a Hot and Cold World." *Maribo* (blog), May 7, 2013. http://simondonner.blogspot.no/2013/05/talking-about-climate-change-in-hot-and.html.

Donner, Simon D., and Jeremy McDaniels. "The Influence of National Temperature Fluctuations on Opinions About Climate Change in the US Since 1990." *Climatic Change* 118, no. 3–4 (February 5, 2013): 537–50. doi:10.1007/s10584-012-0690-3.

Doran, Peter T., and Maggie Kendall Zimmerman. "Examining the Scientific Consensus on Climate Change." *Eos, Transactions American Geophysical Union* 90, no. 3 (2009): 22. doi:10.1029/2009EO030002.

Dreyfus, Hubert L. *Being-in-the-World: A Commentary on Heidegger's Being and Time, Division I.* Cambridge, MA: MIT Press, 1991.

Dunlap, Riley E. "The New Environmental Paradigm Scale: From Marginality to Worldwide Use." *Journal of Environmental Education* 40, no. 1 (September 2008): 3–18. doi:10.3200/JOEE.40.1.3-18.

ecoAmerica. *Climate Impacts: Take Care and Prepare.* Washington, DC: ecoAmerica and Lake Research Partners, August 2012. http://ecoamerica.org/wp-content/uploads/2012/12/ecoAmericaClimateImpactsReportAug2012a.pdf.

———. *New Facts, Old Myths: Environmental Polling Trends.* Washington, DC: ecoAmerica, Summer 2013. http://ecoamerica.org/wp-content/uploads/reports/ecoAmerica_TrendReport_Summer13.pdf.

*Economist.* "Plenty of Gloom" (editorial). *Economist*, December 18, 1997. http://www.economist.com/node/455855.

———. "Climate Change: Cold Comfort" (editorial). *Economist*, June 16, 2012. http://www.economist.com/node/21556805.

———. "Daily Chart: Danger of Death!" *Economist*, February 14, 2013. http://www.economist.com/blogs/graphicdetail/2013/02/daily-chart-7.

———. "Commodity Prices in the (Very) Long Run." *Economist*, March 12, 2013. http://www.economist.com/blogs/freeexchange/2013/03/resource-prices.

———. "Unburnable Fuel: Energy Firms and Climate Change." *Economist*, May 4, 2013. http://www.economist.com/news/business/21577097-either-governments-are-not-serious-about-climate-change-or-fossil-fuel-firms-are.

*EcoWatch.* "Police Close 'Climategate' Investigation into Hacked Emails with Mystery Unsolved." *EcoWatch*, July 19, 2012. http://ecowatch.com/2012/07/19/climategate.

Egebark, Johan, and Mathias Ekström. *Can Indifference Make the World Greener?*, IFN Working Paper No. 975, 2013. http://www.ifn.se/wfiles/wp/wp975.pdf.

Eliot, T. S. *Four Quartets.* New York: Harcourt Brace Jovanovich, 1971.

Fawcett, Tina. "Personal Carbon Trading: A Policy Ahead of Its Time?" *Energy Policy* 38, no. 11 (November 2010): 6868–76. doi:10.1016/j.enpol.2010.07.001.

Feldman, Lauren, Teresa A. Myers, Jay D. Hmielowski, and Anthony Leiserowitz. "The Mutual Reinforcement of Media Selectivity and Effects: Testing the Reinforcing Spirals Framework in the Context of Global Warming." *Journal of Communication* 64, no. 4 (August 2014): 590–611. doi:10.1111/jcom.12108.

Ferraro, Paul J., and Michael K. Price. "Using Non-Pecuniary Strategies to Influence Behavior: Evidence from a Large Scale Field Experiment." National Bureau of Economic Research (NBER) Working Paper Series. NBER, July 2011.

Festinger, Leon. *A Theory of Cognitive Dissonance.* Stanford, CA: Stanford University Press, 1962.

Festinger, Leon, and James M. Carlsmith. "Cognitive Consequences of Forced Compliance." *Journal of Abnormal and Social Psychology* 58, no. 2 (1959): 203–10. doi:10.1037/h0041593.

Festinger, Leon, Stanley Schachter, and Henry W. Riecken. *When Prophecy Fails.* London: Pinter & Martin, 2008.

Fischer, Barry. "Clever Campaign in Denmark: If You Save Energy, You Get Free Pizza." *Outlier* (Opower company blog), December 13, 2012. http://blog.opower.com /2012/12/clever-campaign-in-denmark-if-you-save-energy-you-get-free-pizza.

Fischhoff, B. "Hot Air: The Psychology of CO-Induced Climatic Change." In J. Harvey (ed.), *Cognition, Social Behavior and the Environment.* Hillsdale, NJ: Erlbaum, 1981, 163–84.

Fisher, Andy. *Radical Ecopsychology: Psychology in the Service of Life.* 2nd ed. SUNY Series in Radical Social and Political Theory. Albany: State University of New York Press, 2013.

Food and Agriculture Organization (FAO). *Food Wastage Footprint: Impacts on Natural Resources: Summary Report.* FAO, 2013. www.fao.org/docrep/018/i3347e/i3347e.pdf, accessed October 20, 2014.

Foucault, Michel. *History of Madness.* New York: Routledge, 2006.

Franklin, Benjamin. *The Autobiography of Benjamin Franklin.* Dover Thrift Editions. New York: Dover Publications, 1996.

Freud, Sigmund. *Civilization and Its Discontents.* Translated by James Strachey. New York: W. W. Norton, 2010.

Futerra Sustainability Communications. *Sell the Sizzle.* Futerra Sustainability Communications, 2012. http://www.futerra.co.uk/downloads/Sellthesizzle.pdf.

G8+5 Academies. *G8+5 Academies' Joint Statement: Climate Change and the Transformation of Energy Technologies for a Low Carbon Future.* Washington, DC: The National Academies, May 2009. http://www.nationalacademies.org/includes /G8+5energy-climate09.pdf.

Gardiner, S. M., and L. Hartzell-Nichols. "Ethics and Global Climate Change." *Nature Education* 3, no. 10 (2012): 5.

Gibbons, John. "Time to Sign up for the Climate Change War." *Irishtimes,* December 4, 2013. http://www.irishtimes.com/news/environment/time-to-sign-up-for-the -climate-change-war-1.1616915?mode=print&ot=example.AjaxPageLayout.ot.

Gibson, James William. *A Reenchanted World: The Quest for a New Kinship with Nature.* 1st ed. New York: Metropolitan Books, 2009.

Gilbert, Daniel. "If Only Gay Sex Caused Global Warming." *Los Angeles Times,* July 2006. http://articles.latimes.com/2006/jul/02/opinion/op-gilbert2.

Gilens, Martin, and Benjamin Page. "Testing Theories of American Politics: Elites, Interest Groups, and Average Citizens." *Perspectives on Politics,* in press (2014).

Global Commission on the Economy and Climate. *The New Climate Economy Report.* New Climate Economy (website), June 9, 2014. http://newclimateeconomy.report.

Global e-Sustainability Initiative (GeSI). *GeSI SMARTer2020: The Role of ICT in Driving a Sustainable Future*. GeSI (website). http://gesi.org/SMARTer2020.

Goldstein, Noah J., Robert B. Cialdini, and Vladas Griskevicius. "A Room with a Viewpoint: Using Social Norms to Motivate Environmental Conservation in Hotels." *Journal of Consumer Research* 35, no. 3 (October 2008): 472–82. doi:10.1086/586910.

Greaker, Mads, Per Espen Stoknes, Knut H. Alfsen, and Torgeir Ericson. "A Kantian Approach to Sustainable Development Indicators for Climate Change." *Ecological Economics* 91 (July 2013): 10–18. doi:10.1016/j.ecolecon.2013.03.011.

Greenpeace. "Help Save Santa's Home This Christmas." Greenpeace UK blog, December 5, 2013. http://www.greenpeace.org.uk/blog/climate/help-save-santa%E2%80%99s -home-christmas-20131204.

———. *Clicking Clean: How Companies Are Creating the Green Internet*. Greenpeace, April 2014. http://www.greenpeace.org/usa/global/usa/planet3/pdfs/clickingclean.pdf.

Griskevicius, Vladas, Stepanie M. Cantú, and Mark Vugt. "The Evolutionary Bases for Sustainable Behavior: Implications for Marketing, Policy, and Social Entrepreneurship." *Journal of Public Policy & Marketing* 31, no. 1 (2012): 115–28.

Griskevicius, Vladas, and Mark Vugt. "Let's Use Evolution to Turn Us Green." *New Scientist*, September 21, 2012. http://www.newscientist.com/article/mg21528820 .200-lets-use-evolution-to-turn-us-green.html?full=true&print=true.

Groves, Martha. "Mountain Lion Makes Itself at Home in Griffith Park." *Los Angeles Times*, August 14, 2012. http://articles.latimes.com/2012/aug/14/local /la-me-griffith-park-mountain-lion-20120814.

Haberl, H., K.-H. Erb, and F. Krausmann. "Global Human Appropriation of Net Primary Production (HANPP)." *Encyclopedia of Earth*, September 3, 2013. http://www .eoearth.org/view/article/153031.

Hallet, Emily, and Julie O'Brien. "It's Not All About the Money: New Research Finds the Cost of Energy Isn't the Only Thing Utility Customers Care About." *Our Thinking* (Opower company blog), October 11, 2013. http://blog.opower.com/2013/10/its -not-all-about-the-money-new-research-finds-the-cost-of-energy-isnt-the-only -thing-utility-customers-care-about.

Hamblin, James. "If You Have Allergies or Asthma, Talk to Your Doctor About Cap and Trade." *Atlantic*, August 26, 2014. http://www.theatlantic.com/health/archive/2014 /08/asthma-and-climate-policy/379119.

Hamilton, Clive. "Why We Resist the Truth About Climate Change." Paper presented at the Climate Controversies: Science and Politics conference, Museum of Natural Sciences, Brussels, October 2010. http://clivehamilton.com/pdfdownloader.php?pdf =http://clivehamilton.com/wp-content/uploads/2012/11/why_we_resist _the_truth_about_climate_change.pdf&title=Why%20We%20Resist%20the%20 Truth%20About%20Climate%20Change.

———. "What History Can Teach Us About Climate Change Denial." In *Engaging with Climate Change: Psychoanalytic and Interdisciplinary Perspectives*, edited by Sally Weintrobe. The New Library of Psychoanalysis. Abingdon, UK, and New York: Routledge, 2012.

Hamilton, Clive, and Tim Kasser. "Psychological Adaptation to the Threats and Stresses of a Four Degree World." Paper presented at the Four Degrees and Beyond conference, Oxford University, September 2009. http://www.eci.ox.ac.uk/4degrees/ppt /poster-hamilton.pdf.

Hamilton, Jo. "Keeping up with the Joneses in the Great British Refurb: The Impacts and Limits of Social Learning in Eco-Renovation," in *Engaging the Public with Climate Change: Behaviour Change and Communication*, edited by Lorraine Whitmarsh, Saffron O'Neill, and Irene Lorenzoni. London , Washington, DC: Earthscan, 2011.

Hamilton, Lawrence. *Do You Trust Scientists About the Environment?* Carsey School of Public Policy at the Scholars' Repository, May 1, 2014. Paper 213. http://scholars.unh .edu/carsey/213.

Harbert, Nancy. "In Despair Over the Polar Bear." *Time*, August 17, 2007. http://content .time.com/time/health/article/0,8599,1654087,00.html.

Hardin, Garrett. "The Tragedy of the Commons," *Science* 162, no. 3859 (December 13, 1968): 1243–48. doi:10.1126/science.162.3859.1243.

Harding, Stephan. *Animate Earth: Science, Intuition and Gaia.* White River Junction, VT: Chelsea Green, 2006.

Hardisty, D. J., E. J. Johnson, and E. U. Weber. "A Dirty Word or a Dirty World?: Attribute Framing, Political Affiliation, and Query Theory." *Psychological Science* 21, no. 1 (December 10, 2009): 86–92. doi:10.1177/0956797609355572.

Harré, Niki. *Psychology for a Better World: Strategies to Inspire Sustainability.* Auckland, NZ: Department of Psychology, University of Auckland, 2011. psych.auckland.ac.nz /psychologyforabetterworld.

Harrison, Paul. *Elements of Pantheism.* 2nd ed. Coral Springs, FL: Llumina Press, 2004.

Hart, P. S., and E. C. Nisbet. "Boomerang Effects in Science Communication: How Motivated Reasoning and Identity Cues Amplify Opinion Polarization About Climate Mitigation Policies." *Communication Research* 39, no. 6 (August 11, 2011): 701–23. doi:10.1177/0093650211416646.

Hartcher, Peter. "Cool Heads Missing in the Pressure Cooker." *Sydney Morning Herald*, April 6, 2007. http://www.smh.com.au/news/environment/cool-heads-missing-in -the-pressure-cooker/2007/04/05/1175366406378.html?page=fullpage#contentSwap1.

Harvey, John H., ed. *Cognition, Social Behavior, and the Environment.* Hillsdale, NJ: L. Erlbaum, 1981.

Haukeland, Per Ingvar. *Dyp Glede: Med Arne Næss Inn I Dypøkologien.* 1. utg. Oslo: Flux, 2008.

Hawken, Paul. *Blessed Unrest: How the Largest Social Movement in History Is Restoring Grace, Justice and Beauty to the World.* n.p.: Penguin, 2007.

Hawken, Paul, Amory B. Lovins, and L. Hunter Lovins. *Natural Capitalism: Creating the Next Industrial Revolution.* Boston: Little, Brown, 1999.

———. *Natural Capitalism: The Next Industrial Revolution.* 10th anniversary ed. London: Earthscan, 2010.

Heck, Stefan, Matt Rogers, and Paul Carroll. *Resource Revolution: How to Capture the Biggest Business Opportunity in a Century.* Seattle: Amazon Publishing, 2014.

Heidegger, Martin. *Basic Writings: From Being and Time (1927) to The Task of Thinking (1964)*. Rev. and expanded ed. New York: Harper Perennial Modern Thought, 2008.

Hellevik, Ottar. *Ipsos MMI's Survey "Norsk Monitor."* Personal communication. MMI, April 10, 2014.

Hernes, Gudmund. *Hot Topic—Cold Comfort: Climate Change and Attitude Change*. Oslo: NordForsk, 2012.

Hillman, James. *City & Soul*. Edited by Robert J. Leaver. 1st ed. *Uniform Edition of the Writings of James Hillman*, vol. 2. Putnam, CT: Spring Publications, 2006.

———. *Alchemical Psychology*. 1st ed. *Uniform Edition of the Writings of James Hillman*, vol. 5. Putnam, CT: Spring Publications, 2010.

———. *Re-Visioning Psychology*. New York: HarperPerennial, 1992.

———. *The Thought of the Heart and The Soul of the World*. Dallas: Spring Publications, 1992.

———. *We've Had a Hundred Years of Psychotherapy—and the World's Getting Worse*. 1st ed. San Francisco, CA: HarperSanFrancisco, 1992.

———. *The Soul's Code: In Search of Character and Calling*. New York: Warner Books, 1997.

Hohle, Sigrid M. "'Nudging' Sustainable Food Choices: The Role of Defaults, Frames, Habits and Nature Relatedness." Master's thesis, University of Oslo, Institute of Psychology, 2014.

Hoofnagle, Mark. "About." *Denialism Blog*, April 30, 2007. http://scienceblogs.com /denialism/about.

———. "What Is at the Root of Denial? A Must Read from Chris Mooney in *Mother Jones*." *Denialism Blog*, June 6, 2013. http://scienceblogs.com/denialism/2013/06/06 /what-is-at-the-root-of-denial-a-must-read-from-chris-mooney-in-mother-jones.

Hopey, Don. "Climate Change to Boost Health Problems." *Pittsburgh Post-Gazette*, May 25, 2014. http://www.post-gazette.com/news/environment/2014/05/26/Climate -change-to-boost-health-problems/stories/201405260078.

Hopkins, Rob. "Paul Kingsnorth on Living with Climate Change." Transition Network (website), March 1, 2014. http://www.transitionnetwork.org/blogs/rob-hopkins /2014-03/paul-kingsnorth-living-climate-change.

Howard, Lee. "Rapid Climate Changes More Deadly than Asteroid Impacts in Earth's Past—Study Shows." *Skeptical Science*, May 27, 2014. http://www.skepticalscience .com/Rapid-climate-change-deadlier-than-asteroid-impacts.html.

Høydal, Håkon. "Hvem Bryr Seg." *VG Helg*, December 1, 2012.

Hulme, Mike. *Why We Disagree About Climate Change: Understanding Controversy, Inaction and Opportunity*. Cambridge, UK, and New York: Cambridge University Press, 2009.

Hunt, Patrick. "Interpreting the Nielsen Study . . . or Why Indonesia Is Waaaaay More Concerned About Global Warming than America." *Shelton Insights* (blog), August 31, 2011. http://sheltoninsights.com/green/interpreting-the-nielsen-study-or-why -indonesia-is-waaaaay-more-concerned-about-global-warming-than-america.

Hunt, Terry L., and Carl P. Lipo. *The Statues That Walked: Unraveling the Mystery of Easter Island*. Berkeley, CA: Counter Point Press, 2012.

Hurlstone, Mark, Stephan Lewandowsky, Ben Newel, and Brittany Sewell. "Curbing Emissions: Framing and Normative Messages Influence $CO_2$ Abatement Policy Preferences." *PLOS ONE*, in press (2014).

Inhofe, James. *The Science of Climate Change: Senate Floor Statement by US Sen. James M. Inhofe (R-Okla), Chairman, Committee on Environment and Public Works*, July 28, 2003. http://www.epw.senate.gov/speechitem.cfm?party=rep&id=230594.

Intergovernmental Panel on Climate Change (IPCC). *Fifth Assessment Report: Synthesis Report*. November 2014. http://www.ipcc.ch.

Intergovernmental Panel on Climate Change (IPCC) Working Group 1. *Climate Change 2013: The Physical Science Basis*. November 27, 2013. http://www.ipcc.ch, accessed October 1, 2014.

Intergovernmental Panel on Climate Change (IPCC) Working Group 3. *Climate Change 2013: Mitigation of Climate Change: Summary for Policymakers*. April 12, 2014. http://www.ipcc.ch.

International New York Times Editorial Board. "The Disappearing Moose." *International New York Times*, October 16, 2013. http://www.nytimes.com/2013/10/17/opinion /the-disappearing-moose.html.

Jacks, David S. "From Boom to Bust: A Typology of Real Commodity Prices in the Long Run." NBER Working Paper No. 18874. National Bureau of Economic Research (NBER), March 2013. http://www.nber.org/papers/w18874, accessed October 1, 2014.

Jacques, Peter J., Riley E. Dunlap, and Mark Freeman. "The Organisation of Denial: Conservative Think Tanks and Environmental Scepticism." *Environmental Politics* 17, no. 3 (June 2008): 349–85. doi:10.1080/09644010802055576.

Janis, Irving L. *Groupthink: Psychological Studies of Policy Decisions and Fiascoes*, 2nd ed. Boston: Houghton Mifflin, 1982.

Jasanoff, S. "A New Climate for Society." *Theory, Culture & Society* 27, no. 2–3 (May 24, 2010): 233–53. doi:10.1177/0263276409361497.

Jenkins, Jack. "Pope Francis Makes Biblical Case for Addressing Climate Change." *ClimateProgress*, May 21, 2014. http://thinkprogress.org/climate/2014/05/21/3440075 /pope-francis-if-we-destroy-creation-creation-will-destroy-us.

Jensen, Derrick. "Beyond Hope." *Orion Magazine*, 2006. http://www.orionmagazine.org /index.php/articles/article/170.

Johnson, E. J., and Daniel Goldstein. "Do Defaults Save Lives?" *Science* 302, no. 5649 (November 21, 2003): 1338–39. doi:10.1126/science.1091721.

Johnson, Eric J., Suzanne B. Shu, Benedict G. C. Dellaert, Craig Fox, Daniel G. Goldstein, Gerald Häubl, Richard P. Larrick, et al. "Beyond Nudges: Tools of a Choice Architecture." *Marketing Letters* 23, no. 2 (May 25, 2012): 487–504. doi:10.1007/ s11002-012-9186-1.

Jung, C. G. *The Collected Works of C. G. Jung*. New York: Pantheon Books, 1953.

Kahan, Dan M. "Fixing the Communications Failure." *Nature* 463, no. 7279 (January 21, 2010): 296–97. doi:10.1038/463296a.

———. "Making Climate-Science Communication Evidence-Based—All the Way Down." *SSRN Electronic Journal*, 2013. doi:10.2139/ssrn.2216469.

———. "Climate Science Communication and the Measurement Problem." *Advances in Political Psychology*, in press (2014).

Kahan, Dan M., Donald Braman, John Gastil, Paul Slovic, and C. K. Mertz. "Culture and Identity-Protective Cognition: Explaining the White Male Effect in Risk Perception." *Journal of Empirical Legal Studies* 4, no. 3 (November 2007): 465–505.

Kahan, Dan M., Hank Jenkins-Smith, and Donald Braman. "Cultural Cognition of Scientific Consensus." *Journal of Risk Research* 14, no. 2 (February 2011): 147–74. doi: 10.1080/13669877.2010.511246.

Kahan, Dan M., Maggie Wittlin, Ellen Peters, Paul Slovic, Lisa Larrimore Ouellette, Donald Braman, and Gregory N. Mandel. "The Tragedy of the Risk-Perception Commons: Culture Conflict, Rationality Conflict, and Climate Change." *SSRN Electronic Journal*, 2011. doi:10.2139/ssrn.1871503.

Kahn, Brian. "Hurricane Sandy Hasn't Shifted Climate Narrative." *Climate Central*, October 29, 2013. http://www.climatecentral.org/news/sandy-didnt-change-a-thing-16669.

Kahn, Peter H., and Patricia H. Hasbach, eds. *Ecopsychology: Science, Totems, and the Technological Species*. Cambridge, MA: MIT Press, 2012.

Kahneman, Daniel. *Thinking, Fast and Slow*. 1st paperback ed. New York: Farrar, Straus and Giroux, 2013.

Kahneman, Daniel, Ed Diener, and Norbert Schwarz. *Well-Being: The Foundations of Hedonic Psychology*. New York: Russell Sage Foundation, 2003.

Kahneman, Daniel, Jack L. Knetsch, and Richard H. Thaler. "Experimental Tests of the Endowment Effect and the Coase Theorem." *Journal of Political Economy* 98, no. 6 (January 1990): 1325. doi:10.1086/261737.

Kallbekken, Steffen, and Håkon Sælen. "'Nudging' Hotel Guests to Reduce Food Waste as a Win–Win Environmental Measure." *Economics Letters* 119, no. 3 (June 2013): 325–27. doi:10.1016/j.econlet.2013.03.019.

Kallbekken, Steffen, Håkon Sælen, and Erlend A. T. Hermansen. "Bridging the Energy Efficiency Gap: A Field Experiment on Lifetime Energy Costs and Household Appliances." *Journal of Consumer Policy* 36, no. 1 (October 2, 2012): 1–16. doi:10.1007/s10603-012-9211-z.

Kant, Immanuel. *Grounding for the Metaphysics of Morals*, 3rd ed. Translated by James W. Ellington. Cambridge, MA: Hackett, 1993 (originally published in 1785).

Keukens, Rob, and Robert van Voren. "Coercion in Psychiatry: Still an Instrument of Political Misuse?" *BMC Psychiatry* 7, suppl. 1 (2007). doi:10.1186/1471-244X-7-S1-S4.

Keynes, J. M. "Economic Possibilities for Our Grandchildren." In J. M. Keynes, *Essays in Persuasion*. London: Macmillan, 1931.

King, Jeff. "A Critique of Western Psychology from an American Indian Psychologist." Edited by Nancy Cater. *Spring: A Journal of Archetype and Culture* 87 (Summer 2012): 37–60.

Klein, Ezra. "How Politics Makes Us Stupid." *Vox*, April 6, 2014. http://www.vox.com/2014/4/6/5556462/brain-dead-how-politics-makes-us-stupid.

Krause, Andrew, and George Basile. "Can Millennials and Social Networking Lead Us to a Sustainable Future?" *GreenBiz*, May 20, 2014. http://www.greenbiz.com/blog /2014/05/20/can-millennials-and-social-networking-lead-us-sustainable-future.

Krausmann, F., K.-H. Erb, S. Gingrich, H. Haberl, A. Bondeau, V. Gaube, C. Lauk, C. Plutzar, and T. D. Searchinger. "Global Human Appropriation of Net Primary Production Doubled in the 20th Century." *Proceedings of the National Academy of Sciences* 110, no. 25 (June 18, 2013): 10324–29. doi:10.1073/pnas.1211349110.

Kroh, Kiley. "Push to Impose Extra Fees on Solar Customers Draws Outrage in Wisconsin." *ThinkProgress*, September 14, 2014. http://thinkprogress.org/climate /2014/09/14/3567244/utility-fees-end-wisconsin-solar.

Krygsman, K., and ecoAmerica. *Campaigns II: Recent Learnings from Other Social Movements*. Washington, DC: ecoAmerica, 2014.

Kübler-Ross, Elisabeth, and David Kessler. *On Grief and Grieving: Finding the Meaning of Grief Through the Five Stages of Loss*. New York: Scribner, 2007.

Kuhn, Thomas S. *The Structure of Scientific Revolutions*. 3rd ed. Chicago: University of Chicago Press, 1996.

Lakoff, George. *Don't Think of an Elephant!: Know Your Values and Frame the Debate: The Essential Guide for Progressives*. White River Junction, VT: Chelsea Green, 2004.

———. "Why It Matters How We Frame the Environment." *Environmental Communication: A Journal of Nature and Culture* 4, no. 1 (March 2010): 70–81. doi:10.1080/17524030903529749.

———. *The All New Don't Think of an Elephant!: Know Your Values and Frame the Debate*. White River Junction, VT: Chelsea Green, 2014.

Latour, Bruno. *We Have Never Been Modern*. Cambridge, MA: Harvard University Press, 1993.

———. "Waiting for Gaia. Composing the Common World Through Arts and Politics." Lecture presented at the launching of SPEAP (the Sciences Po program in arts & politics), the French Institute, London, November 2011.

Leaton, James. *Unburnable Carbon 2013: Wasted Capital and Stranded Assets*. Carbon Tracker, 2013. http://www.carbontracker.org/wp-content/uploads/2014/09/ Unburnable-Carbon-2-Web-Version.pdf.

Leber, Rebecca. "76 Percent of Religious Americans Want a Global Pact Cutting Pollution, Viewing It in Moral and Religious Terms." *ThinkProgress*, December 9, 2011. http://thinkprogress.org/green/2011/12/09/386330/76-percent-of-religious-americans -want-a-global-pact-cutting-pollution-viewing-it-in-moral-and-religious-terms.

Legates, David R., Willie Soon, William M. Briggs, and Christopher Monckton of Brenchley. "Climate Consensus and 'Misinformation': A Rejoinder to Agnotology, Scientific Consensus, and the Teaching and Learning of Climate Change." *Science & Education*, August 30, 2013. doi:10.1007/s11191-013-9647-9.

Leiserowitz, Anthony. "American Risk Perceptions: Is Climate Change Dangerous?" *Risk Analysis* 25, no. 6 (December 2005): 1433–42. doi:10.1111/j.1540-6261.2005.00690.x.

Leiserowitz, Anthony, G. Feinberg, S. Rosenthal, N. Smith, A. Anderson, C. Rose-Renouf, and E. Maibach. *What's in a Name? Global Warming vs. Climate Change*. New Haven, CT: Yale Project on Climate Change Communication, Yale University and George Mason University, 2014.

Leiserowitz, Anthony, E. Maibach, and Connie Roser-Renouf. *Saving Energy at Home and on the Road*. George Mason University: Center for Climate Change Communication, 2009.

Leiserowitz, Anthony, E. Maibach, C. Roser-Renouf, G. Feinberg, and S. Rosenthal. *Climate Change in the American Mind, April 2014*. Yale Project on Climate Change Communication. New Haven, CT: Yale University and George Mason University, 2014.

———. *Politics & Global Warming, Spring 2014*. Yale Project on Climate Change Communication. New Haven, CT: Yale University and George Mason University, 2014.

Leiserowitz, Anthony, N. Smith, and J. R. Marlon. *Americans' Knowledge of Climate Change*. Yale Project on Climate Change Communication. New Haven, CT: Yale University, 2010.

Leopold, Aldo. *A Sand County Almanac: With Essays on Conservation from Round River*. New York: Ballantine Books, 1970.

Lertzman, Renee. "The Myth of Apathy: Psychoanalytic Explorations of Environmental Subjectivity." Chapter 6 in *Engaging with Climate Change: Psychoanalytic and Interdisciplinary Perspectives*, edited by Sally Weintrobe. The New Library of Psychoanalysis. Abingdon, UK, and New York: Routledge, 2012.

Lewandowsky, Stephan, John Cook, Klaus Oberauer, and Michael Marriott. "Recursive Fury: Conspiracist Ideation in the Blogosphere in Response to Research on Conspiracist Ideation." *Frontiers in Psychology* 4, no. 73 (February 2, 2013). doi:10.3389/fpsyg.2013.00073.

Lewandowsky, Stephan, Gilles E. Gignac, and Samuel Vaughan. "The Pivotal Role of Perceived Scientific Consensus in Acceptance of Science." *Nature Climate Change* 3, no. 4 (October 28, 2012): 399–404. doi:10.1038/nclimate1720.

Lewandowsky, Stephan, K. Oberauer, and G. E. Gignac. "NASA Faked the Moon Landing—Therefore, (Climate) Science Is a Hoax: An Anatomy of the Motivated Rejection of Science." *Psychological Science* 24, no. 5 (March 26, 2013): 622–33. doi:10.1177/0956797612457686.

Li, Y., E. J. Johnson, and L. Zaval. "Local Warming: Daily Temperature Change Influences Belief in Global Warming." *Psychological Science* 22, no. 4 (March 3, 2011): 454–59. doi:10.1177/0956797611400913.

Lifton, Robert Jay. "Beyond Psychic Numbing: A Call to Awareness." *American Journal of Orthopsychiatry* 52, no. 4 (1982): 619–29. doi:10.1111/j.1939-0025.1982.tb01451.x.

———. "The Climate Swerve." *New York Times*, August 23, 2014. http://www.nytimes.com/2014/08/24/opinion/sunday/the-climate-swerve.html?_r=1.

Lindzey, Gardner, and Elliot Aronson (eds.). *Handbook of Social Psychology*. 3rd ed. New York and Hillsdale, NJ: Random House. Distributed exclusively by L. Erlbaum Associates, 1985.

Loewenstein, George F., Elke U. Weber, Christopher K. Hsee, and Ned Welch. "Risk as Feelings." *Psychological Bulletin* 127, no. 2 (2001): 267–86. doi:10.1037//0033-2909.127.2.267.

Lorenzoni, Irene, Sophie Nicholson-Cole, and Lorraine Whitmarsh. "Barriers Perceived to Engaging with Climate Change Among the UK Public and Their Policy Implications." *Global Environmental Change* 17, no. 3–4 (August 2007): 445–59. doi:10.1016/j.gloenvcha.2007.01.004.

Lovelock, James. *Gaia: A New Look at Life on Earth.* Oxford, UK, and New York: Oxford University Press, 2000.

Lovelock, James E., and Lynn Margulis. "Atmospheric Homeostasis by and for the Biosphere: The Gaia Hypothesis." *Tellus XXVI,* no. 1–2 (1974).

Lovins, Amory. *Reinventing Fire: Bold Business Solutions for the New Energy Era.* White River Junction, VT: Chelsea Green, 2013.

Lucas, Robert E. *Lectures on Economic Growth.* Cambridge, MA: Harvard University Press, 2002.

Luntz, Frank. *The Environment: A Cleaner Safer, Healthier America* (Frank Luntz Memorandum to Bush White House). Luntz Research, 2002. https://www2.bc.edu /~plater/Newpublicsite06/suppmats/02.6.pdf.

MacRae, Paul. "Why Climate Science Is a Textbook Example of Groupthink." *Watts Up with That,* May 1, 2012. http://wattsupwiththat.com/2012/04/30/why-climate -science-is-a-textbook-example-of-groupthink.

Macy, Joanna. *Active Hope: How to Face the Mess We're in Without Going Crazy.* Novato, CA: New World Library, 2012.

Maibach, Edward, Teresa Myers, and Anthony Leiserowitz. "Climate Scientists Need to Set the Record Straight: There Is a Scientific Consensus That Human-Caused Climate Change Is Happening." *Earth's Future* 2, no. 5 (May 2014). doi:10.1002/2013EF000226.

Maibach, Edward W., Matthew Nisbet, Paula Baldwin, Karen Akerlof, and Guoqing Diao. "Reframing Climate Change as a Public Health Issue: An Exploratory Study of Public Reactions." *BMC Public Health* 10, no. 1 (2010): 299. doi:10.1186/1471-2458-10-299.

Manne, Robert. "How Vested Interests Defeated Climate Science: A Dark Victory." *The Monthly,* 2012. http://www.themonthly.com.au/issue/2012/august/1344299325/robert -manne/dark-victory.

Manning, Mark. "The Effects of Subjective Norms on Behaviour in the Theory of Planned Behaviour: A Meta-Analysis." *British Journal of Social Psychology* 48, no. 4 (December 2009): 649–705. doi:10.1348/014466608X393136.

Marshall, George. *Don't Even Think About It: Why Our Brains Are Wired to Ignore Climate Change.* Bloomsbury USA, 2014.

Marshall, George. *How to Engage Your Community and Communicate About Climate Change.* Climate Outreach Information Network (COIN), 2011.

Mayer, F. Stephan, and Cynthia McPherson Frantz. "The Connectedness to Nature Scale: A Measure of Individuals' Feeling in Community with Nature." *Journal of Environmental Psychology* 24, no. 4 (December 2004): 503–15. doi:10.1016/j. jenvp.2004.10.001.

McCarthy, Cormac. *The Road.* New York: Vintage Books, 2006.

McCright, Aaron M., and Riley E. Dunlap. "Anti-Reflexivity: The American Conservative Movement's Success in Undermining Climate Science and Policy." *Theory, Culture & Society* 27, no. 2–3 (March 1, 2010): 100–33. doi:10.1177/0263276409356001.

———. "Cool Dudes: The Denial of Climate Change Among Conservative White Males in the United States." *Global Environmental Change* 21, no. 4 (October 2011): 1163–72. doi:10.1016/j.gloenvcha.2011.06.003.

————. "The Politicization of Climate Change and Polarization in the American Public's Views of Global Warming 2001–2010." *Sociological Quarterly* 52, no. 2 (March 2011): 155–94. doi:10.1111/j.1533-8525.2011.01198.x.

McDonough, William. *The Upcycle*. 1st ed. New York: North Point Press, a division of Farrar, Straus and Giroux, 2013.

McEvoy, John G. *The Historiography of the Chemical Revolution: Patterns of Interpretation in the History of Science*. London and Brookfield, VT: Pickering & Chatto, 2010.

McKewon, Elaine. "The Use of Neoliberal Think Tank Fantasy Themes to Delegitimise Scientific Knowledge of Climate Change in Australian Newspapers." *Journalism Studies* 13, no. 2 (April 2012): 277–97. doi:10.1080/1461670X.2011.646403.

McKibben, Bill. "Global Warming's Terrifying New Math." *Rolling Stone*, October 5, 2012.

————. "Growing Up Global: The Common Future of Rich and Poor." *PBS Wide Angle*, August 29, 2002. http://www.pbs.org/wnet/wideangle/episodes/growing-up-global /the-common-future-of-rich-and-poor/3112.

McKinsey & Company. *Impact of the Financial Crisis on Carbon Economics: Version 2.1 of the Global Greenhouse Gas Abatement Cost Curve*. McKinsey & Company, 2010. http://www.mckinsey.com/client_service/sustainability/latest_thinking /greenhouse_gas_abatement_cost_curves.

McKinsey Global Institute. *Resource Revolution: Tracking Global Commodity Markets*. McKinsey & Company, 2013. http:// www.mckinsey.com/insights/energy_resources_materials/ resource_revolution_tracking_global_commodity_markets.

McLean, Pam. "Buckinghamshire County Council Rolls Out Carbon Conversations." *Carbon Conversations*, March 13, 2014. http://carbonconversations.org/news/2014 /buckinghamshire-county-council-rolls-out-carbon-conversations.

McNeley, James Kale. *Holy Wind in Navajo Philosophy*. Tucson: University of Arizona Press, 1981.

Meador, Jonathan. "Kentucky Lawmakers Attack Climate Change Science in Discussion on Carbon Regulations." WFPL News, July 4, 2014. http://wfpl.org/post/kentucky -lawmakers-attack-climate-change-science-discussion-carbon-regulations.

Merleau-Ponty, Maurice. *Phenomenology of Perception*. Translated by Donald A. Landes, London: Routledge, 2013.

Michaels, Patrick. "Why Hasn't The Earth Warmed in Nearly 15 Years?" *Forbes*, July 15, 2011. http://www.forbes.com/sites/patrickmichaels/2011/07/15/ why-hasnt-the-earth-warmed-in-nearly-15-years.

Monbiot, George. "Climate Change Enlightenment Was Fun While It Lasted. But Now It's Dead." *The Guardian*, September 20, 2010. http://www.theguardian.com /commentisfree/2010/sep/20/climate-change-negotiations-failure.

————. *Feral: Searching for Enchantment on the Frontiers of Rewilding*. London: Allen Lane, 2013.

Mooney, Chris. "Conspiracy Theorists Are More Likely to Doubt Climate Science." *Mother Jones*, June 6, 2013. http://www.motherjones.com/blue-marble/2013/06 /conspiracy-theorists-also-doubt-climate-science.

———. "Conservatives Don't Deny Climate Science Because They're Ignorant. They Deny It Because of Who They Are." *Mother Jones*, June 26, 2014. http://www.motherjones .com/environment/2014/06/dan-kahan-climate-change-ideology-scientific-illiteracy.

———. "The Strange Relationship Between Global Warming Denial and . . . Speaking English." *Mother Jones*, July 22, 2014. http://www.motherjones.com /environment/2014/07/climate-denial-us-uk-australia-canada-english.

Moore, Thomas. *Care of the Soul: A Guide for Cultivating Depth and Sacredness in Everyday Life*. New York: HarperCollins, 1992.

Moran, Alan. "Alarm on Global Warming Just a Load of Hot Air." *Institute of Public Affairs Australia*, August 9, 2006. http://www.ipa.org.au/news/1223/alarm-on -global-warming-just-a-load-of-hot-air/category/4.

Morgan, M. Granger, ed. *Risk Communication: A Mental Models Approach*. Cambridge, UK, and New York: Cambridge University Press, 2002.

Moser, Susanne C., and Lisa Dilling. *Creating a Climate for Change: Communicating Climate Change and Facilitating Social Change*. Cambridge, UK, and New York: Cambridge University Press, 2007.

Muller, Richard. "The Conversion of a Climate-Change Skeptic." *New York Times*, July 28, 2012. http://www.nytimes.com/2012/07/30/opinion/the-conversion-of-a-climate -change-skeptic.html?pagewanted=all&_r=1&.

Myers, Teresa A., Matthew C. Nisbet, Edward W. Maibach, and Anthony A. Leiserowitz. "A Public Health Frame Arouses Hopeful Emotions About Climate Change: A Letter." *Climatic Change* 113, no. 3–4 (June 28, 2012): 1105–12. doi:10.1007 /s10584-012-0513-6.

Næss, Arne. *Scepticism*. International Library of Philosophy and Scientific Method. London and New York: Routledge and Kegan Paul, Humanities Press, 1968.

———. *Ecology, Community, and Lifestyle: Outline of an Ecosophy*. Translated by David Rothenberg. Cambridge, UK, and New York: Cambridge University Press, 1989.

National Oceanic and Atmospheric Administration (NOAA). "NOAA's Ten Signs of a Warming World." NOAA, 2009. http://cpo.noaa.gov/warmingworld.

National Public Radio (NPR), "A Christian Climate Scientist's Mission to Convert Nonbelievers." *The Sunday Conversation*, June 8, 2014. http://www.npr.org/2014/06 /08/319831143/climate-scientist-climate-change-is-a-christian-issue-too.

National Safety Council (NSC). "Odds of Dying." NSC, 2013. http://www.nsc.org/ news_resources/injury_and_death_statistics/Pages/TheOddsofDyingFrom.aspx.

Newell, Ben R., and Andrew J. Pitman. "The Psychology of Global Warming: Improving the Fit Between the Science and the Message." *Bulletin of the American Meteorological Society* 91, no. 8 (August 2010): 1003–14. doi:10.1175/2010BAMS2957.1.

*New Scientist*. "Climate Change: Awaking the Sleeping Giants." *New Scientist*, February 12, 2005. http://www.newscientist.com/article/mg18524864.400-climate-change -awaking-the-sleeping-giants.html.

Nielsen Company. "Sustainability Survey: Global Warming Cools Off as Top Concern." *Nielsen Press Room*, August 28, 2011. http://www.nielsen.com/us/en/press-room /2011/global-warming-cools-off-as-top-concern.html.

———. *Sustainable Efforts and Environmental Concerns Around the World*, August 2011.

Nisbet, Matthew C. "Communicating Climate Change: Why Frames Matter for Public
    Engagement." *Environment: Science and Policy for Sustainable Development*, March–
    April 2009. http://www.environmentmagazine.org/Archives/Back%20Issues
    /March-April%202009/Nisbet-full.html.

Nisbet, M. C., and T. Myers. "The Polls—Trends: Twenty Years of Public Opinion About
    Global Warming." *Public Opinion Quarterly* 71, no. 3 (August 11, 2007): 444–70.
    doi:10.1093/poq/nfm031.

Nisbet, M. C., and D. A. Scheufele. "What's Next for Science Communication? Promising
    Directions and Lingering Distractions." *American Journal of Botany* 96, no. 10
    (October 1, 2009): 1767–78. doi:10.3732/ajb.0900041.

NOAA National Climatic Data Center (NCDC). "Global Analysis—Annual 2013." NCDC,
    2014. http://www.ncdc.noaa.gov/sotc/global/2013/13.

Noel, Daniel C. *The Soul of Shamanism: Western Fantasies, Imaginal Realities*. New York:
    Continuum, 1999.

Nolan, J. M., P. W. Schultz, R. B. Cialdini, N. J. Goldstein, and V. Griskevicius. "Normative
    Social Influence Is Underdetected." *Personality and Social Psychology Bulletin* 34, no.
    7 (May 9, 2008): 913–23. doi:10.1177/0146167208316691.

Nordhaus, William D. *A Question of Balance: Weighing the Options on Global Warming
    Policies*. New Haven, CT: Yale University Press, 2008.

Norgaard, Kari Marie. *Living in Denial: Climate Change, Emotions, and Everyday Life*.
    Cambridge, MA: MIT Press, 2011.

Norgaard, Kari M., and Ron Reed. "Emotional Impacts of Environmental Decline: What
    Can Native Cosmologies Teach Sociology About Race, Emotions and Environmental
    Justice?" *Journal of Health and Social Behavior* 55, in press (2014).

Norton, David W. "Constructing 'Climategate' and Tracking Chatter in an Age of Web
    n.0." Center for Social Media, AUSOC School of Communications, American
    University, Washington, DC, September 2010.

Nuccitelli, Dana. "Can a Carbon Tax Work Without Hurting the Economy? Ask British
    Columbia." *The Guardian*, July 30, 2013. http://www.theguardian.com/environment
    /climate-consensus-97-per-cent/2013/jul/30 climate-change-british-columbia
    -carbon-tax.

———. "Let's Be Honest—The Global Warming Debate Isn't About Science."
    *The Guardian*, October 4, 2013. http://www.theguardian.com/environment
    /climate-consensus-97-per-cent/2013/oct/04/global-warming-debate-not-about
    -science.

Nye, M., and J. Burgess. *Promoting Durable Change in Household Waste and Energy Use
    Behaviour*. Department for Environment, Food & Rural Affairs, UK, February 2008.
    http://storage.globalcitizen.net/data/topic/knowledge/uploads/2011061415321
    5533.pdf.

Olli, E., G. Grendstad, and D. Wollebaek. "Correlates of Environmental Behaviors:
    Bringing Back Social Context." *Environment and Behavior* 33, no. 2 (March 1, 2001):
    181–208. doi:10.1177/0013916501332002.

Olson, Randy. *Don't Be Such a Scientist: Talking Substance in an Age of Style*. Washington,
    DC: Island Press, 2009.

O'Neill, S., and S. Nicholson-Cole. "'Fear Won't Do It': Promoting Positive Engagement with Climate Change Through Visual and Iconic Representations." *Science Communication* 30, no. 3 (January 7, 2009): 355–79. doi:10.1177/1075547008329201.

Oreskes, Naomi, and Erik M. Conway. *Merchants of Doubt: How a Handful of Scientists Obscured the Truth on Issues from Tobacco Smoke to Global Warming.* 1st US ed. New York: Bloomsbury Press, 2010.

Organization for Economic Cooperation and Development (OECD). *Green Growth Indicators 2014.* OECD Green Growth Studies. OECD Publishing, 2014. http://www .oecd-ilibrary.org/environment/green-growth-indicators-2013_9789264202030-en.

Ornstein, Robert E., and Paul R. Ehrlich. *New World New Mind: Moving Toward Conscious Evolution.* Cambridge, MA: Malor Books, 2000.

Ostrom, E. "Revisiting the Commons: Local Lessons, Global Challenges." *Science* 284, no. 5412 (April 9, 1999): 278–82. doi:10.1126/science.284.5412.278.

Painter, James. *Climate Change in the Media: Reporting Risk and Uncertainty.* London: I. B. Tauris, 2013.

Palmer, Martin. *Faith in Conservation: New Approaches to Religions and the Environment.* Directions in Development. Washington, DC: World Bank, 2003.

Paris, Max. "Outdoor Rink Climate Change Project Gets Hundreds of Citizen Scientists." CBC News, January 23, 2013. http://www.cbc.ca/news/politics/outdoor -rink-climate-change-project-gets-hundreds-of-citizen-scientists-1.1344016.

Pash, Barbara. "Coalition Brings Solar Energy Co-ops to Baltimore Area." *Baltimore Sun,* July 14, 2014. http://www.baltimoresun.com/news/maryland/baltimore-county /towson/ph-tt-solar-energy-co-op-0730-20140725,0,7624406.story.

Patz, Jonathan A., Howard Frumkin, Tracey Holloway, Daniel J. Vimont, and Andrew Haines. "Climate Change Challenges and Opportunities for Global Health." *Journal of the American Medical Association,* September 22, 2014. doi:10.1001/ jama.2014.13186.

Pauli, Gunter A. *The Blue Economy: 10 Years, 100 Innovations, 100 Million Jobs.* Taos, NM: Paradigm Publications, 2010.

Paulson, Henry. "The Coming Climate Crash." *New York Times,* June 21, 2014. http://www.nytimes.com/2014/06/22/opinion/sunday/lessons-for-climate-change -in-the-2008-recession.html?_r=1.

Pew Research Center. *Public Priorities: Deficit Rising, Terrorism Slipping.* Pew Research Center, January 23, 2012. www.pewglobal.org.

———. *More Say There Is Solid Evidence for Global Warming.* Pew Research Center, October 15, 2012. http://www.people-press.org/files/legacy-pdf/10-15-12%20 Global%20Warming%20Release.pdf.

———. *Climate Change and Financial Instability Seen as Top Global Threats.* Pew Research Center, June 24, 2013. www.pewglobal.org.

———. *Deficit Reduction Declines as Policy Priority.* Pew Research Center, January 27, 2014. www.pewglobal.org.

Phillips, Ari. "New Texas GOP Platform Calls on Politicians to Ignore Climate Change." *Climate Progress,* June 5, 2014. http://thinkprogress.org/climate/2014/06/05/3445339 /texas-gop-2014-platform-climate-change.

Pichert, Daniel, and Konstantinos V. Katsikopoulos. "Green Defaults: Information Presentation and Pro-Environmental Behaviour." *Journal of Environmental Psychology* 28, no. 1 (March 2008): 63–73. doi:10.1016/j.jenvp.2007.09.004.

Pidgeon, Nick. "Public Understanding Of, and Attitudes To, Climate Change: UK and International Perspectives and Policy." *Climate Policy* 12, sup01 (September 2012): S85–106. doi:10.1080/14693062.2012.702982.

Piketty, Thomas. *Capital in the Twenty-First Century*. Cambridge MA: The Belknap Press of Harvard University Press, 2014.

Plumer, Brad. "Two Degrees: How the World Failed on Climate Change." *Vox*, April 22, 2014. http://www.vox.com/2014/4/22/5551004/two-degrees.

Poortinga, Wouter, Alexa Spence, Lorraine Whitmarsh, Stuart Capstick, and Nick F. Pidgeon. "Uncertain Climate: An Investigation into Public Scepticism About Anthropogenic Climate Change." *Global Environmental Change* 21, no. 3 (August 2011): 1015–24. doi:10.1016/j.gloenvcha.2011.03.001.

Poortinga, Wouter, Lorraine Whitmarsh, and Christine Suffolk. "The Introduction of a Single-Use Carrier Bag Charge in Wales: Attitude Change and Behavioural Spillover Effects." *Journal of Environmental Psychology* 36 (December 2013): 240–47. doi:10.1016/j.jenvp.2013.09.001.

Porritt, Jonathon. *The World We Made: Alex McKay's Story from 2050*. London: Phaidon Press, 2013.

Powell, James L. "Methodology." James Lawrence Powell: Science and Global Warming (website), February 1, 2014. http://www.jamespowell.org/methodology/method.html.

Randall, Alex, Jo Salsbury, and Zach White. *Moving Stories: The Voices of People Who Move in the Context of Environmental Change*. Climate Outreach Information Network (COIN), 2014.

Randall, Rosemary. "Loss and Climate Change: The Cost of Parallel Narratives." *Ecopsychology* 1, no. 3 (September 2009): 118–29. doi:10.1089/eco.2009.0034.

Randers, Jorgen. *2052: A Global Forecast for the Next Forty Years*. White River Junction, VT: Chelsea Green, 2012.

———. "Greenhouse Gas Emissions per Unit of Value Added ('GEVA')—A Corporate Guide to Voluntary Climate Action." *Energy Policy* 48 (September 2012): 46–55. doi:10.1016/j.enpol.2012.04.041.

Revkin, Andrew. "On 'Unburnable Carbon' and the Specter of a 'Carbon Bubble.'" *New York Times*, May 3, 2013. http://dotearth.blogs.nytimes.com/2013/05/03/on -unburnable-carbon-and-the-specter-of-a-carbon-bubble/?ref=globalwarming.

Riffkin, Rebecca. "Climate Change Not a Top Worry in US." Gallup News Service, March 12, 2014. http://www.gallup.com/poll/167843/climate-change-not-top-worry.aspx.

Ritter, Malcolm. "Climate Psychology 101: Global Warming a Tough Sell for Human Psyche." *Cleveland.com* (blog), December 17, 2009. http://blog.cleveland.com /world_impact/print.html?entry=/2009/12/climate_psychology_101_global.html.

Rockström, Johan, Will Steffen, Kevin Noone, Åsa Persson, F. Stuart Chapin, Eric F. Lambin, Timothy M. Lenton, et al. "A Safe Operating Space for Humanity." *Nature* 461, no. 7263 (September 24, 2009): 472–75. doi:10.1038/461472a.

Ropeik, David. *Risk: A Practical Guide for Deciding What's Really Safe and What's Dangerous in the World Around You.* Boston: Houghton Mifflin, 2002.

Rosenberg, Tina. *Join the Club: How Peer Pressure Can Transform the World.* New York: W. W. Norton, 2011.

Rosentrater, L. D., I. Saelensminde, F. Ekstrom, G. Bohm, A. Bostrom, D. Hanss, and R. E. O'Connor. "Efficacy Trade-Offs in Individuals' Support for Climate Change Policies." *Environment and Behavior* 45, no. 8 (July 17, 2012): 935–70. doi:10.1177/0013916512450510.

Roszak, Theodore, Mary E. Gomes, and Allen D. Kanner, eds. *Ecopsychology: Restoring the Earth, Healing the Mind.* San Francisco: Sierra Club Books, 1995.

Royal Society and National Academy of Sciences. *Climate Change: Evidence and Causes.* PDF Booklet report, Kindle ed., March 11, 2014. http://dels.nas.edu/resources/static -assets/exec-office-other/climate-change-full.pdf.

Running, Steve. "Five Stages of Climate Grief." Numerical Terradynamic Simulation Group (website), November 26, 2007. http://www.ntsg.umt.edu/files/5StagesClimateGrief.htm.

Rust, Mary-Jayne, and Nick Totten. *Vital Signs: Psychological Responses to Ecological Crisis.* London and Herndon, VA: Karnac Books and Stylus Publishing (distributor), 2012.

Scherer, Glenn. "Climate Science Predictions Prove Too Conservative." *Scientific American*, December 6, 2012. http://www.scientificamerican.com/article/climate -science-predictions-prove-too-conservative.

Schultz, P. W. "Changing Behavior with Normative Feedback Interventions: A Field Experiment of Curbside Recycling." *Basic and Applied Social Psychology* 21 (1999): 25–36.

Schultz, P. Wesley, Chris Shriver, Jennifer J. Tabanico, and Azar M. Khazian. "Implicit Connections with Nature." *Journal of Environmental Psychology* 24, no. 1 (March 2004): 31–42. doi:10.1016/S0272-4944(03)00022-7.

Shaw, Martin. *A Branch from the Lightning Tree: Ecstatic Myth and the Grace in Wildness.* Ashland, OR: White Cloud Press, 2011.

Sinaceur, Marwan, Chip Heath, and Steve Cole. "Emotional and Deliberative Reactions to a Public Crisis: Mad Cow Disease in France." *Psychological Science* 16, no. 3 (March 2005): 247–54. doi:10.1111/j.0956-7976.2005.00811.x.

Skogstad, Knut. "Tesla Model S: Nå Knuser Den Alle Konkurrentene." *tv2.no.* September 12, 2013. http://www.tv2.no/underholdning/broom/tesla-model-s-naa-knuser -den-alle-konkurrentene-4120063.html.

Smith, Daniel B. "Is There an Ecological Unconscious?" *New York Times*, January 31, 2010.

Smith, Nicholas, and Anthony Leiserowitz. "The Rise of Global Warming Skepticism: Exploring Affective Image Associations in the United States Over Time." *Risk Analysis* 32, no. 6 (June 2012): 1021–32. doi:10.1111/j.1539-6924.2012.01801.x.

Spence, Alexa, and Nick Pidgeon. "Framing and Communicating Climate Change: The Effects of Distance and Outcome Frame Manipulations." *Global Environmental Change* 20, no. 4 (October 2010): 656–67. doi:10.1016/j.gloenvcha.2010.07.002.

Spence, Alexa, Wouter Poortinga, and Nick Pidgeon. "The Psychological Distance of Climate Change." *Risk Analysis* 32, no. 6 (June 2012): 957–72. doi:10.1111/j.1539-6924.2011.01695.x.

Stephenson, N. L., A. J. Das, R. Condit, S. E. Russo, P. J. Baker, N. G. Beckman, D. A. Coomes, et al. "Rate of Tree Carbon Accumulation Increases Continuously with Tree Size." *Nature* 507, no. 7490 (January 15, 2014): 90–93. doi:10.1038/nature12914.

Sterman, John D. "Risk Communication on Climate: Mental Models and Mass Balance— Supporting Online Material." *Science* 322, no. 5901 (October 2008). http://www .sciencemag.org/content/322/5901/532/rel-suppl/5213589f9e4f5151/suppl/DC1.

Stern, Paul C. "New Environmental Theories: Toward a Coherent Theory of Environmentally Significant Behavior." *Journal of Social Issues* 56, no. 3 (January 2000): 407–24. doi:10.1111/0022-4537.00175.

———. "Psychology: Fear and Hope in Climate Messages." *Nature Climate Change* 2, no. 8 (June 17, 2012): 572–73. doi:10.1038/nclimate1610.

Stiglitz, Joseph. "Progress, What Progress?" *OECD Observer* 272, March 2009.

Stiglitz, Joseph, Amartya Sen, and Jean-Paul Fitoussi. *Report of the Commission on the Measurement of Economic Performance and Social Progress (CMEPSP)*. CMEPSP, 2009. http://www.stiglitz-sen-fitoussi.fr/documents/rapport_anglais.pdf.

Stoknes, Per Espen. "Økopsykologi: Fra et intrapsykisk til et økologisk Selv." Dissertation for the degree of psychologist, University of Oslo, 1994.

———. *Sjelens Landskap: Refleksjoner over Natur Og Myter*. Oslo: Cappelen, 1996.

———. *Money & Soul: The Psychology of Money and the Transformation of Capitalism*. Totnes, UK: Green Books, 2009.

———. "Rethinking Climate Communications and the 'Psychological Climate Paradox.'" *Energy Research & Social Science* 1 (March 2014): 161–70. doi:10.1016/j. erss.2014.03.007.

Stoknes, Per Espen, and Frede Hermansen (eds.). *Lær Av Fremtiden*. Gyldendal akademisk, 2004.

Stoll-Kleemann, S., Tim O'Riordan, and Carlo C. Jaeger. "The Psychology of Denial Concerning Climate Mitigation Measures: Evidence from Swiss Focus Groups." *Global Environmental Change*, 2001.

Story Group. *National Climate Assessment Videos*. http://thestorygroup.org/category /nationalclimateassessment, accessed October 1, 2014.

Sturgis, Patrick, and Nick Allum. "Science in Society: Re-Evaluating the Deficit Model of Public Attitudes." *Public Understanding of Science* 13, no. 1 (January 1, 2004): 55–74. doi:10.1177/0963662504042690.

Sunstein, Cass. *Risk and Reason: Safety, Law, and the Environment*. Cambridge, UK, and New York: Cambridge University Press, 2002.

———. "Behavioral Economics, Consumption, and Environmental Protection." In forthcoming handbook on research in sustainable consumption, preliminary draft, 2013.

*SustainableBusiness.com*. "US Solar Industry Employs More than Coal, Gas Industries Combined." *SustainableBusiness.com*, January 28, 2014. http://www.sustainablebusiness.com/index.cfm/go/news.display/id/25475.

Sustainia. "Sustainable Solutions." Sustainia (website), 2014. http://www.sustainia.me/solutions.

Swift, Art. "Americans Again Pick Environment Over Economic Growth." Gallup News Service, March 9, 2014. http://www.gallup.com/poll/168017/americans -again-pick-environment-economic-growth.aspx.

Tam, Kim-Pong. "Concepts and Measures Related to Connection to Nature: Similarities and Differences." *Journal of Environmental Psychology* 34 (June 2013): 64–78. doi:10.1016/j.jenvp.2013.01.004.

Tavris, Carol, and Elliot Aronson. *Mistakes Were Made (but Not by Me): Why We Justify Foolish Beliefs, Bad Decisions, and Hurtful Acts*. Orlando, FL: Harcourt, 2008.

Thaler, Richard H. *Nudge: Improving Decisions About Health, Wealth, and Happiness*. New Haven, CT: Yale University Press, 2008.

Theel, Shauna. "Patrick Michaels: Cato's Climate Expert Has History of Getting It Wrong." *Skeptical Science*, July 13, 2013. http://skepticalscience.com/patrick-michaels -history-getting-climate-wrong.html.

Tinjum, Aaron. "Surging Forward: Opower Surpasses 4 Billionth Kilowatt Hour of Energy Savings." *Our Thinking* (Opower company blog), February 18, 2014. http://blog .opower.com/2014/02/surging-forward-opower-surpasses-4-billionth-kilowatt -hour-of-energy-savings.

Treuer, G. A., E. U. Weber, K. C. Appelt, A. E. Goll, and R. D. Crookes. "Weathering the Storm: Status Quo Adjustments Explain Successful Policy Implementation." *Political Behavior*, in press.

Truelove, H. B., A. R. Carrico, E. U. Weber, K. T. Raimi, and M. P Vandenbergh. "Positive and Negative Spillover of Pro-Environmental Behavior." *Global Environmental Change*, in press (2014).

Tversky, Amos, and Daniel Kahneman. "Rational Choice and the Framing of Decisions." *Journal of Business* 59, no. 4 (October 1, 1986): S251–78. doi:10.2307/2352759.

———. "Loss Aversion in Riskless Choice: A Reference-Dependent Model." *Quarterly Journal of Economics* 106, no. 4 (November 1, 1991): 1039–61. doi:10.2307/2937956.

UK Energy Research Centre. *Low Carbon Jobs: The Evidence for Net Job Creation from Policy Support for Energy Efficiency and Renewable Energy*. November 10, 2014. http://www.edie.net/.

Union of Concerned Scientists. *Cooler Smarter: Practical Steps for Low-Carbon Living*. Edited by Seth Shulman. Washington, DC: Island Press, 2012.

UNEP, 2011, *Towards a Green Economy: Pathways to Sustainable Development and Poverty Eradication—A Synthesis for Policy Makers*, www.unep.org/greeneconomy.

United Nations University and United Nations Environment Program (UNEP). *Inclusive Wealth Report 2012: Measuring Progress Toward Sustainability*. Cambridge, UK: Cambridge University Press, 2012.

US Green Building Council (USGBC). "About USGBC: Better Buildings Are Our Legacy." USGBC (website), 2014. http://www.usgbc.org/About.

US House of Representatives Committee on Energy and Commerce Minority Staff. *The Anti-Environment Record of the US House of Representatives 112th Congress*. June 18, 2012. http://democrats.energycommerce.house.gov/sites/default/files/documents/ Anti-Environment-Voting-Record-of-112th-Congress-Summary-Final.pdf.

Valkeapää, Nils-Aslak. *Trekways of the Wind*. Translated by Ralph Salisbury and Lars Nordström. Kautokeino. DAT publishers, 1994.

Van der Linden, S. "Exploring Beliefs About Bottled Water and Intentions to Reduce Consumption: The Dual-Effect of Social Norm Activation and

Persuasive Information." *Environment and Behavior*, December 20, 2013. doi:10.1177/0013916513515239.

Van der Post, Laurens. *The Heart of the Hunter*. Harmondsworth, UK: Penguin Books, in association with Hogarth Press, 1965.

Vigen, Michelle, and Susan Mazur-Stommen. "Reaching the 'High-Hanging Fruit' Through Behavior Change: How Community-Based Social Marketing Puts Energy Savings Within Reach." American Council for an Energy-Efficient Economy (ACEEE) White Paper, October 2012. http://www.aceee.org/files/pdf/white-paper /high-hanging-fruit-cbsm.pdf.

Von Wright, Georg Henrik. *Explanation and Understanding*. London: Routledge, 2011.

Wackernagel, Mathis. "Global Footprint Network: Ecological Footprint." Global Footprint Network (website), August 4, 2014. http://www.footprintnetwork.org.

*Washington Times*. "Chilling Climate-Change News" (editorial). *Washington Times*, December 18, 2012. http://www.washingtontimes.com/news/2012/dec/18/chilling -climate-change-news.

Weber, Elke U. "Experience-Based and Description-Based Perceptions of Long-Term Risk: Why Global Warming Does Not Scare Us (Yet)." *Climatic Change* 77, no. 1–2 (July 21, 2006): 103–20. doi:10.1007/s10584-006-9060-3.

Weber, Elke U., Eric J. Johnson, Lisa Zaval, Elizabeth Keenan, and Ye Li. "Heuristics and Constructed Beliefs in Climate Change Perception: Effect of Outdoor Temperature, Question Construction, and Cognitive Primes." Center for Research on Environmental Decisions, June 25, 2013. http://cred.columbia.edu/research /all-projects/heuristics-and-constructed-beliefs-in-climate-change-perception -effect-of-outdoor-temperature-question-construction-and-cognitive-primes.

Wei, Max, Shana Patadia, and Daniel M. Kammen. "Putting Renewables and Energy Efficiency to Work: How Many Jobs Can the Clean Energy Industry Generate in the US?" *Energy Policy* 38, no. 2 (February 2010): 919–31. doi:10.1016/j.enpol.2009.10.044.

Weintrobe, Sally, ed. *Engaging with Climate Change: Psychoanalytic and Interdisciplinary Perspectives*. The New Library of Psychoanalysis. Abingdon, UK, and New York: Routledge, 2012.

Weissbecker, Inka, ed. *Climate Change and Human Well-Being: Global Challenges and Opportunities*. International and Cultural Psychology. New York: Springer, 2011.

Weitzman, Martin L. "On Modeling and Interpreting the Economics of Catastrophic Climate Change." *Review of Economics and Statistics* 91, no. 1 (February 2009): 1–19. doi:10.1162/rest.91.1.1.

———. "Fat-Tailed Uncertainty in the Economics of Catastrophic Climate Change." *Review of Environmental Economics and Policy* 5, no. 2 (June 1, 2011): 275–92. doi:10.1093/reep/rer006.

———. "GHG Targets as Insurance Against Catastrophic Climate Damages." *Journal of Public Economic Theory* 14, no. 2 (March 2012): 221–44. doi:10.1111/j.1467-9779.2011.01539.x.

Weizsäcker, Ernst U. von. *Factor Five: Transforming the Global Economy Through 80% Improvements in Resource Productivity: A Report to the Club of Rome*. London and Sterling, VA: Earthscan/The Natural Edge Project, 2009.

White, L. "The Historical Roots of Our Ecologic Crisis." *Science* 155, no. 3767 (March 10, 1967): 1203–07. doi:10.1126/science.155.3767.1203.

Whitmarsh, Lorraine. "Behavioural Responses to Climate Change: Asymmetry of Intentions and Impacts." *Journal of Environmental Psychology* 29, no. 1 (March 2009): 13–23. doi:10.1016/j.jenvp.2008.05.003.

———. "Scepticism and Uncertainty About Climate Change: Dimensions, Determinants and Change Over Time." *Global Environmental Change* 21, no. 2 (2011): 690–700.

Whitmarsh, Lorraine, Saffron O'Neill, and Irene Lorenzoni, eds. *Engaging the Public with Climate Change: Behaviour Change and Communication.* London and Washington, DC: Earthscan, 2011.

Wilcove, David, and Steve Curwood. "Messed Up Migrations." *Living on Earth,* December 6, 2013. http://www.loe.org/shows/segments.html?programID=13-P13-00049 &segmentID=4.

Wildcat, Daniel R. *Red Alert!: Saving the Planet with Indigenous Knowledge.* Speaker's Corner. Golden, CO: Fulcrum, 2009.

Wilson, Adrian. *Coal Blooded.* National Association for the Advancement of Colored People, Baltimore, 2011.

Wilson, Edward O. *Biophilia.* Cambridge, MA: Harvard University Press, 1984.

Wilson, M., M. Daly, and S. Gordon. "The Evolved Psychological Apparatus of Human Decision-Making Is One Source of Environmental Problems." In *Behavioral Ecology and Conservation Biology,* edited by T. M. Caro. New York: Oxford University Press, 1998.

Wolf, Johanna, and Susanne C. Moser. "Individual Understandings, Perceptions, and Engagement with Climate Change: Insights from In-Depth Studies Across the World." *Wiley Interdisciplinary Reviews: Climate Change* 2, no. 4 (July 2011): 547–69. doi:10.1002/wcc.120.

Worden, J. William. *Grief Counseling and Grief Therapy: A Handbook for the Mental Health Practitioner.* 4th ed. New York: Springer, 2009.

World Business Council for Sustainable Development (WBCSD). *Vision 2050: The New Agenda for Business.* WBCSD (website), 2011. http://www.wbcsd.org/vision2050.aspx.

World Commission on Environment and Development. *Our Common Future* (Brundtland Report). Oxford, UK: Oxford University Press, 1987.

World Wildlife Fund (WWF), CDP, and McKinsey & Company. *The 3% Solution: Driving Profits Through Carbon Reduction.* WWF (website), 2013. http://www.worldwildlife .org/projects/the-3-solution.

Wysham, Daphne. "The Six Stages of Climate Grief." *Huff Post Green,* September 4, 2012. http://www.huffingtonpost.com/daphne-wysham/the-six-stages-of-climate _b_1852425.html.

# —INDEX—

Abbott, Anthony John, 52
Abram, David, 165, 181, 207, 213
Abramowitz, Yosef, 143
abstraction, climate change as, 40, 45
acceptance, as grief stage, 185–86
active denial, 16–17
active optimism, 220–21
active skepticism, 221–22
active worlding, 203
adaptation, as approach to climate change, 117
affective component, attitude, 57, 60
African Americans, 174
air, 7, 165–70, 196, 199–216, 224, 226–27
    active worlding, 203
    components of, 166, 204
    connectedness to, 186, 203, 211, 213–16
    as earth's skin, 166–68
    historic roots to sense of living air, 211–12
    personifying, 203
    *poiesis* of, 204–8
    psyche as, 205–10
air pollution. *See* pollution
Alarmed, the, 58–59, 65–66, 105
Albrecht, Glenn, 174
allergies, 114–15
Alliance to Save Energy, 154
alternative climate initiatives, 93
alternative energy. *See* clean energy
altruism, competitive, 30
altruistic values, 198–99
Amundsen, Per-Willy, 11
ancestral forces ("old mind"), 27–29, 34, 41
anger, 115–16, 179–80, 185
animals, 134, 190
    extinction, 171, 186
    human bond with, 181–82, 186
Ansoff, Igor, 152

Antarctica, 40, 107, 218
anthropocentrism, 177
anti-climate movement, 25–26, 48–50, 78, 83, 118. *See also* denialism
apathy. *See* indifference
apocalyptic stories, x, 132–33, 139, 147, 149
Apple
    HomeKit, 128
    renewable energy, commitment to, 138
appliances, 126, 128–29
apps
    peer behavior and, 98–100
    smart-home nudges, 129
Arctic, 87, 146
    distancing of, 29, 40
    permafrost/ice, melting of, xviii, 14–15, 107, 169, 194–95
    wind currents, 169
Aristotle, 204–5
Asch, Solomon, 30–31
Asia, 6, 42, 142–43. *See also* China
asthma, 114–15, 174
attic insulation program, 126
attitudes, climate change, 57–67
    behavior and (*See* behavior)
    cognitive dissonance and, 60–67
    components of, 57–60, 58–60, 104
    explicit and implicit, 58
    polarized, 72, 83, 90, 107
Australia, 4, 14, 52, 174
Austria, 124
Avaaz, 107
awareness, 186, 206–7

Bandura, Albert, 103
bargaining, 185
Barton, Joe, 12
behavior
    attitudes and, 57, 58, 66–67, 82, 130

eco-networks, pro-environmental
behavior of, 105
new behavior, components needed
for, 129
nudging approach (*See* nudging)
peer, 95–100
behavioral economics, 129–30
Being-in-the-world, 210
Bell, Larry, 11–12
Benestad, Rasmus, 56–57
Benyus, Janine, 133
Berkeley Earth, 3
Bernstein, Jerome, 181–83
Berry, Wendell, 175
bikes, 98–99, 115, 137
biomimicry, 133
biophilia, 188, 199
biospheric values, 198–99
*Blessed Unrest,* 106
Bloomberg, Michael, 113
Bogotá, Colombia, 125
Borderland, living in, 181–86
Breakthrough Institute, 119
breath (breathing), 166–67, 203–6, 211,
213, 224, 226
British Columbia, 121
Burgess, Michael, 12
Bush, George H. W., 7
businesses, 134
climate-change solutions, role in,
21–22, 136–38
signals for, 152

Camus, Albert, 225–26
Carbon Conversations, 104
carbon dioxide, 36–38, 40, 48, 110, 166,
203. *See also* greenhouse gas emissions
communicating levels of, 42
ethical duty to cut emissions of, 158–60
health, effect on, 115
meat farming, produced by, 128
natural sinks for, 136
value added, relationship of emissions
to, 153–55

carbon footprint, reducing, 104, 126. *See
also* climate action
carbon offsets, 49, 64, 125, 131
carbon price/tax, xix–xx, 20–21, 75, 93, 111
Australia, 52
British Columbia, 121
carbon or earmarked offset frame *vs.,* 49
resistance to, 96
Cautious, the, 58–60
Chamberlain, Neville, 218
China, 63, 118, 158
Christianity
beliefs and stories, air in, 211–12
good-*versus*-bad dualism, 179
relationship to nature of, 142–44
Cialdini, Robert, 95, 97, 98, 99
cities, 23, 27, 93, 119, 222–23. *See also
specific cities*
climate action in, 102–6, 126–27,
129, 134
green growth, measuring, 154
parking, restricted, 125
re-wilding, 146–47
transport options in, 137
citizen science, 102–3
clean energy, 120–21, 124, 126–27, 137–38.
*See also* solar power; wind power
climate action, xx, 22. *See also* framing
(messaging) climate change; political
action; signals of progress; simple
solutions; social networks
alternative initiatives, 93
barriers to, 81–84, 87–94
ethics and (*See* ethics)
individual responsibility for (*See*
individual responsibility)
media, role of, 97, 102–3
psychology of (*See* psychology)
small-actions approach, 88–89, 91
social citizens, acting as, 91–92
swerve toward, xiv–xv, 23, 64, 78–80,
90, 92, 125
climate change. *See also* greenhouse gas
emissions

distance from (*See* distance from
   climate change)
ethics of (*See* ethics)
framing (messaging) (*See* framing
   (messaging) climate change)
global-warming terminology *vs.*, 47–49
prioritizing, xx, 4–6, 17, 32, 58
urgency of (*See* urgency, climate
   change)
climate change, concern about, xviii–xix, 46
   Climategate, effect of, 50, 65, 75
   cultural identity and, 73–74
   distance from climate change, effect of
      (*See* distance from climate change)
   facts, role of, 35–39, 59–60, 81
   global-warming frame *vs.* climate-
      change frame, 47–49
   percent of population concerned by,
      4–5, 29, 32, 58, 65–66, 73–74
   risk perception and, 39, 42
   social attitudes and, 58–59
   weather events, effect of, 41–42, 80,
      112–13
climate-change denial, xiii–xv, 3, 6–7,
   9–26, 28–29, 82. *See also* anti-climate
   movement; climate contrarians; cultural
   denial; denial; resistance
   accusation, term as, 76, 108–9
   attractiveness of, 25–26
   avoiding, 90, 175
   Climategate, framing of, 49–50
   cognitive dissonance and, 63–64
   corporate, 78
   disaster narrative and, 149
   ethics of, 23–25
   groupthink claim, 7, 13, 31
   identity and, 70–80
   media coverage of, 11–15, 67–68, 118
   nations, role of, 20–21
   optimism as type of, 217–18
   percent of population agreeing with, xv,
      xviii, 3–6, 25
   as resistance, 76–78, 109
   skepticism and, 9–11

supportive evidence, search for, 75–76
swerve in, 78–80
climate-change facts, xviii–xx, 3
   communicating, attempts at, ix–x,
      xix–xx, 3, 59–60, 81, 84, 92–93
   cultural cognition and, 72–74
   meaning of facts, internalizing of, 170
   perception of, 35–39
climate-change skeptics. *See* skepticism
climate contrarians, 10, 11. *See also*
   climate-change denial
climate disruption, 22, 49, 77, 204–6. *See
   also* climate change; symptoms, climate
   disruption
Climate Hawks Vote, 80
climate psychology. *See* psychology
climate science, ix–x, xiv–xv, 15–16. *See
   also* climate-change facts
   communicating (*See* communication)
   consensus on, xviii–xix, 3, 59–60, 101
   diversity of ideas and alternatives, 31
   models, 9
   skepticism, role of (*See* skepticism)
climate symptoms. *See* symptoms, climate
   disruption
Climategate, 49–50, 65, 75
coal, 40, 46, 71, 121, 132, 139, 178
   health risks of, 174
   jobs in coal industry, 121
   natural gas as replacement for, 71
   production costs, 137–38
   waste in production of, 135, 136
coastal countries, 42
cognition, cultural, 72–76
cognitive component, attitude, 57, 58, 60
cognitive dissonance, 60–67, 89, 104
cognitive psychology, 35–53
   climate-change facts perception, 35–39
   climate-change-related events, role of,
      39–42
   climate-change risk perception, 39
   distance from climate change, effect of,
      39–43, 82
   fast and slow systems, 40–42

framing, 46–53
  risk perception, 39, 42–46
Cohen, Stanley, 16, 79
Colbert, Stephen, 133
collectives. *See* social networks
Committee for a Constructive
  Tomorrow, 13
communication, xix, 3, 31, 42. *See
  also* climate-change facts; framing
  (messaging) climate change; signals of
  progress; story-based communications
  barriers to, 34, 74–75, 81–84, 89–90
  climate denial and climate resistance
    distinguished, 77–78
  defeatist, 132–33
  good-*versus*-bad dualism, moralizing
    effect of, 179–80
  of individual responsibility, 88
  IPCC reports, style of, 55–57
  for locality, 107
  numbers in, 42–43, 151, 153, 162
  positive strategies, 90, 92–94, 107
  simple solutions as strategy for, 93,
    129–31
  stronger system, need for, 122–23
  success criteria for, 90
  visualization, 152
community. *See also* social networks
  climate action by, 101–6
  despair, recognition through, 176
  stimulating, 90
competitive altruism, 30
Concerned, the, 58–59, 65–66
concrete industry, 137
confirmation bias, 72–73
connectedness, nature. *See* nature
  connectedness
consciousness, 206
conservatism, 49, 66–69, 72–74, 82, 144
conspiracy, climate change as, xviii, 7, 11,
  64, 75
control
  locus of, 40
  risk perception and, 44, 45

cooling systems, 136–37
Copenhagen climate negotiations (2009),
  49–50, 65, 154
Corner, Adam, 105, 134
corporations. *See* businesses
cost-benefit analysis, 111
costs. *See also* carbon price/tax; energy
  savings; losses, costs, and sacrifice,
  framing climate change as
  as climate-action motivator, 96, 98–100,
    112–14
  smart energy-use meters, displayed
    by, 129
Crompton, Tom, 118–19
cultural awareness, swerve in, 19, 78–80.
  *See also* climate action, swerve toward
cultural cognition, 72–76, 82
cultural denial, 18–20, 22–23, 72–76,
  219–20
  evolutionary psychology and, 28–29
  swerve in, 78–80
cultural identity, 72–76, 82
cultural support for climate-change
  solutions, 22–23, 119, 219–20
culture, 134, 225
  climate symptoms and values, 198–99
  evolutionary psychology and, 27–29, 34
  good-*versus*-bad dualism, 179
Curitiba, Brazil, 125

Dark Mountain Project, 178, 184
Dasein, 210
Dasgupta, Partha, 151, 157
De Geus, Arie, 117
default options, green, x, 124–27
deforestation, 28, 137, 194
delayed gratification, 32
denial, 16–20, 130. *See also* climate-change
  denial; denialism
  active and passive, 16–17
  definition of, 16
  of depression, 177
  as first stage of grief, 185–86
  group cohesion created by, 75

psychological (individual), 16–19,
    22–23
denialism, 18–19. *See also* anti-climate
    movement
    ethics of, 23–25
    funded, 78
    public and political, 92
    social networks, use of, 108–9
    uncertainty, response to, 116–18
    victim and heroes, perception as,
        23–25, 75
Denmark, 128
depression, ix, 171–89
    Borderland, living in, 181–86
    despair paradox, 188–89
    as grief stage, 185–86
    humbleness and, 177–81
    love, value of, 187
    nature and land, bond with, 186–88
despair, 219
    eco-anxieties, mordor of, 173–77
    Environmental Despair, 175
    as grief stage, 185
    opening up through, 176–77
    paradox of, 188–89
destruction, framing climate change as, 5,
    110, 114–16
developing countries, 5–6
Diamond, Jared, 28, 221
Dierker, Chris, 102–3
disaster, framing climate change as, 5, 110,
    149, 174
Disengaged, the, 58–60
disillusionment, 91–92
Dismissive, the, 59, 64, 116. *See also*
    denialism
dissonance, 88, 90, 124, 130. *See also*
    cognitive dissonance; social dissonance
distance from climate change, 39–43, 82
    climate-action motivation and, 98
    closeness of air and, 167–68, 215
    communications and, 90, 107, 114, 130
distance of nature from humans, 195–96,
    199–200. *See also* nature connectedness

diversity, xiv–xv, 93, 141, 147
Doha climate negotiations (2012), 13
doom, 5, 82, 130
Doubtful, the, 59–60, 63, 66, 116
Drucker, Peter, 152

earmarked offsets frame, 49
earth
    finite nature of, 196–97
    self-restorative capabilities of, 224–25
Earth Hour, 103–4
Easter Island, story of, 28
eco-anxieties, mordor of, 173–77
eco-despair. *See* despair
eco-networks, pro-environmental behavior
    of, 105
eco-pathologies, 191–95
eco-philosophy, 194, 207
eco-psychology, 193, 200
eco-symptoms. *See* symptoms, climate
    disruption
ecoAmerica, 117
ecocide, 144–45
ecological footprint assessments, 156–57
ecological mind-set, 196–98
ecological Self (identity), 197–98
ecology, science of, 199
Ecology of Grief, 175
economic development, 119–21, 135–38,
    152, 154, 229
economic incentives, emission reduction,
    130. *See also* carbon price/tax
*Ecosystem Millennium Assessment
    Report,* 161
egoistic values, 198–99
electric cars, 99
Elkjop, 126
emotional component, attitude, 57, 60, 104
emotional response to climate messages,
    115–16, 179
Energidienst company, 127
energy efficiency, 111, 121, 126, 136–38,
    154
energy savings, 153

green nudges for, 128, 129
social networks, motivated by, 98–100,
    103–4, 108
environment. *See* air; nature; pollution
Environmental Despair, 175
environmental risks. *See* risk perception,
    climate change
ethics
    of climate-change denial, 23–25
    as climate-message frame, 116–19
    indicators to measure, 158–60
    stewardship story and, 144–45
Europe, 4–6, 105. *See also specific countries*
European Union, 78, 126
evolutionary psychology, 27–34, 41, 100
explicit attitudes, 58

face-to-face interactions, 100–102
facts, climate change. *See* climate-change
    facts
fear, 13–14, 33, 36
    denial and, 16–17, 82, 90
    framing, effect of, 46, 60, 64, 82, 90
    risk and, 43–45
feedback, 151–53, 194–95, 197
*Feral,* 145
Ferraro, Paul, 97
Festinger, Leon, 61–62
first-loop learning, 193–94
flock status. *See* status
Fogg, B. J., 129
food
    local food production, 104
    meat consumption, reducing, 128
    waste, 127, 137
Fort Collins, Colorado, 127
Fossil-fuel industries
    climate change, response to, 70–71, 74
    communities affected by, 174
    greening of, 21–22
    jobs in, 121
fossil fuels, 22, 46, 71, 132, 218. *See also*
    coal; greenhouse gas emissions; pollution
    energy use, 82, 102, 136

production costs, 137–38, 154
waste in production of, 135–38
framing (messaging) climate change, 3,
    31, 39, 42, 46–53, 108, 110–23, 215. *See
    also* losses, costs, and sacrifice, framing
    climate change as; shaming; story-based
    communications
    anti-climate messages, 25–26, 48–50, 118
    barriers to, xiii, xv, 34, 74–75, 81–84,
        89–90
    carbon tax or earmarked offset, 49
    Climategate or e-mail theft, 49–50, 65
    cognitive dissonance and, 64
    destruction, 5, 110, 114–16
    disaster, 5, 110, 149, 174
    emotional response to, 115–16
    ethics, 116–19
    global warming or climate change, 47–49
    insurance, 111–14
    opportunity, 110–11, 119–21
    preparedness, 116–19
    small-step solutions, 91
    supportive, xx, 90, 93, 122–23, 150, 153
    uncertainty, 110, 116–19
Francis, Pope, 143
free-market energy, 120–21
Freud, Sigmund, 16, 95, 191
Fromm, Erich, 188
Futerra, 139

Galileo, deniers compared to, 75
GDP (gross domestic product), 112, 140,
    151–52, 154, 156–57, 177
Germany, 124, 126–27
Gibbons, John, 179–80
Gibson, James, 147
Gilbert, Daniel, 33
global warming, 47–49. *See also* climate
    change
Goldberg, Noah, 95
good-*versus*-bad dualism, 179
Goodman, Paul, 92
Google
    home technology (Nest), 128

renewable energy, commitment
    to, 138
Gore, Al, 65, 71, 72, 78
government regulation, 71
Great Grief, x, 171–73, 176
green buildings, 121
Green Cities Index, 105
green communities, 104–5
green defaults, 124–27
green-growth narrative, 135–38
green-growth signals, 153–56
green nudges for homes, 126–27
Green Sports Alliance, 105
Greenhouse Emissions Per Value Added
    (GEVA), 154–55
greenhouse gas emissions, xv, xix–xx, 7,
    36–39, 121, 137. *See also* carbon dioxide;
    carbon offsets; carbon price/tax
    action to cut (*See* climate action)
    company value added and, 153–55
    ethical duty to cut, 158–60
    Kuznets curve and, 166
    natural sinks for, 136
    resistance to cutting, 12, 21–22, 51, 59,
        63, 71–72
Greenpeace, 179
Greensburg, Kansas, 105
GreeNudge, 126
Grief, Ecology of, 175
grief, moving beyond, x, 185–86
Griffith, Morgan, 12–13
Griskevicius, Vladas, 97
gross domestic product. *See* GDP (gross
    domestic product)
groups. *See also* social networks
    anti-climate action, 108–9
    climate action by, 101
    conformity of, 95 (*See also* groupthink)
groupthink, xiv–xv, 7, 13, 31, 95
guilt, 58, 82, 88
    as climate-action motivator, 97
    guilt-inducing messages, effect of,
        60–61, 64, 92, 179, 180
    overcoming, 90

Halloween Gone Green, 103
Hamilton, Clive, 221, 226
Hansen, James, 46
happiness, 151
    counting, 156–58
    well-being narrative, 139–42
Happy Planet Index, 157
Harding, Stephan, 171
Harrison, Paul, 204
Hartmann, Thom, 221
Hawken, Paul, 106
Hawthorne effect, 54–55
health issues, 114–15, 174
heat-related sickness and deaths, 114–15
heating systems, 136–37
Hedegaard, Connie, 78
Heidegger, Martin, 210
helplessness, 40, 82, 88, 101
Higgins, Polly, 144
Hillman, James, 171, 176, 190–91, 200, 203
hoax, climate change as, xviii, 7, 11–13, 50,
    64, 141
Hoofnagle, Mark, 18–19
hope, ix–x, 217–27
    climate messages, response to, 115–16
    grounded, 222
    as type of denial, 217–18
    varieties of, 220–22
    as wishful thinking, 184–85, 218
human capital, 157–58
Human Development Index, 151
humbleness, 177–81
Hunt, Patrick, 42
Hurricane Katrina, 71, 80
Hurricane Sandy, 13, 41, 80

ice melt, xviii, 14–15, 107, 169, 194–95, 218
identity
    ecological, 197–98
    psychology of, 70–80, 82
IEA. *See* International Energy Agency (IEA)
imagination, 210, 214–15
imitation, 27–28, 30–31, 95–96, 100, 104.
    *See also* social norms

implicit attitudes, 58
*Inconvenient Truth, An,* 65
indifference, 24–26, 81, 91, 124, 142, 171
indigenous peoples, 28, 106, 173–74, 198, 212–13
individual (psychological) denial, 16–19, 22–23
individual responsibility, 22–23, 95, 223–25
    ethical duty to mitigate greenhouse gas emissions, 159–60
    helplessness and, 88
    political actions replaced by, 91–92, 125
    small-actions approach, 88–89, 91
individuals, place in world of, xx–xxi, 195–299. *See also* air; nature connectedness
industrial revolution, xv, 42–43, 135–36, 139
industry. *See* businesses
inertia, xvii, 108, 197
Inhofe, James, 13
innovation, xvii, 102, 122, 134–38
insulation
    attic insulation, program for, 126
    energy conserved by, 153
insurance, framing climate action as, 111–14
Integrated Wealth Indicator, 157–58
Intergovernmental Panel on Climate Change (IPCC), 221
    2007 report, 65
    communication style of reports, 37–38, 55–57
    denialists, response by, 13–14, 71
    Nobel Prize, 65
international climate-change negotiations, 20, 22, 93
    Climategate, effect of, 49–50
    Copenhagen 2009, 49–50, 65, 154
    Doha 2012, 13
    ethical duty of nations, 158–59
    support for, 143–44
International Energy Agency (IEA), 38
international law, environmental ethics in, 144–45

Internet, energy use by, 138
investors, 113, 136
invisibility of climate change, 40, 45
IPCC. *See* Intergovernmental Panel on Climate Change (IPCC)
Islam, relationship to nature of, 143

Jensen, Derrick, 184, 189, 221
job growth, 111, 217
    climate action, gains from, 119–22, 134, 136, 141, 154, 220
    climate action, losses from, 72
    local issues, emphasis on, 107
    prioritized over climate change, 6
    as signal, 152
Jones, Alex, 14
Judaism, relationship to nature of, 143
Jung, Carl Gustav, 95, 206, 210

Kahan, Dan, 68, 73, 107–8
Kahneman, Daniel, 40, 41, 51, 58
Kant, Immanuel, 119, 158–60
Kantian Climate Policy Indicator, 159
Kaplan, Robert, 152
Karuk Tribe, 173–74
Kayhoe, Katharine, 75
Keeling, Ralph, 37
Keeling curve, 37
Keynes, John Maynard, 140
Kingsforth, Paul, 184
Klein, Ezra, 67–68
Klein, Melanie, 180
Kübler-Ross, Elisabeth, 185
Kuznets curve, 165–66

Lakoff, George, 47, 49, 121–22
land, bond to, 186–88
Larsen, Pamela, 173
Latin America, 6
Leiserowitz, Anthony, 58
Leopold, Aldo, 183–84
Lertzman, Rene, 171
Levinas, Emmanuel, 24
Lewandosky, Stephan, 74

liberalism, 48, 69, 72, 73, 82
libertarianism, 71, 74, 120
Lifton, Robert Jay, 19, 175
lighting, 136
*Limits to Growth, The,* ix
littering, 97
Little Bear, Leroy, 198
*Living in the Borderland,* 181–83
local food production, 104
long-term scenarios, positive and negative
    elements of, 218–19
Los Angeles, California, 125, 147
loss, aversion to, 51, 82, 111–14
losses, costs, and sacrifice, framing climate
    change as, ix–x, 51–53, 82, 110–14,
    118–21, 130, 220. *See also* costs
Lucas, Robert E., 139
Luntz, Frank, 48
Lyons, Oren, 143

Macy, Joanna, 175
Marxism, climate-change theory compared
    to, 11
McCarthy, Cormac, 28
McKinsey & Company, 154
McLeman, Robert, 102–3
McPherson, Guy, 221
meat consumption, reducing, 128
media, climate-change coverage by, 55,
    60, 170
    climate-action motivation, 97, 102–3
    climate-change denial, 11–15, 67–68,
        118
    disaster narrative, 122, 149
    opportunity, climate change as, 110–11
    signals for, 152
    weather events, effect of, 40, 46
mental health, 115, 174
Merleau-Ponty, Maurice, 207, 215
methane, 128, 169
Michaels, Pat, 11
military, risk planning by, xiv, 112–13
Miller, R. L., 79–80
modern mind-set, 195–97

Monbiot, George, 4, 145–46
Monckton, Christopher, 13
morality, 179. *See also* ethics
    as climate-action motivator, 52, 63–64,
        97, 119, 144, 149
    passive denial and, 149
Moran, Alan, 14
Morano, Marc, 13
motivated reasoning, 73
Münster, Germany, 98–99
Musk, Elon, 133, 137
Myers, Teresa, 115
Myrvold, Ulf, 146
myths, ancient, 168–70, 211

narratives. *See* story-based communications
Nasr, Seyyed H., 143
Næss, Arne, 194
National Oceanic and Atmospheric
    Administration (NOAA), 35–36
nations
    climate-change denial and, 20–21
    ethical duty to mitigate greenhouse gas
        emissions, 158–59
    green growth, measuring, 154–56
Native Americans, relationship to nature
    of, 143
natural capital, 157–58
natural gas. *See* fossil fuels
nature
    human bond with (*See* nature
        connectedness)
    listening to, 193–95
    as separate from humans, 195–96,
        199–200
nature connectedness, 186–88, 200,
    213–16, 224–27. *See also* air; Great Grief
    Borderland, the, 181–82
    indigenous peoples, 173–74, 197–98
Nature Index, 160–61
Navajo, Holy Wind in worldview of,
    212–13
negative feelings, 47, 57, 59, 90, 217
Nelson, Richard, 197–98

Neumann, Matt, 120
New Economics Foundation, 157
nitrogen, natural sinks for, 136
NOAA. *See* National Oceanic and
    Atmospheric Administration (NOAA)
Nobel Prize, 65
nongovernmental organizations, 106
nonprofit organizations, 93
Nordhaus, Ted, 119
Norgaard, Kari Marie, 17, 22
North America, 4–6, 121. *See also*
    United States
Norway, 43, 99, 146–47, 222–23
    climate-change concern, 4–5, 41
    climate-change study, 17
    Kantian Climate Policy Indicator for, 159
    Nature Index, 160–61
nudging, 124
    as climate communication, 129–31
    food and water waste, 127–28
    green defaults, 124–27
    green nudges for homes, 126–27
    meat consumption, reducing, 128
    smart homes, 128–29

ocean conveyors, 169
oil. *See* fossil fuels
"old mind" (ancestral forces), 27–29, 34, 41
opinion leaders, 102
Opower, 98–100
opportunity, framing climate change as,
    110–11, 119–21
optimism, 217–21
Oreskes, Naomi, 221
organ donations, as default option, 124
organizational theory, 193–94
Oslo, Norway, 146–47, 222–23
ozone hole, 48

paper conservation, 125
passive denial, 16–17, 149
passive houses, 136–37
passive optimism, 220–21
passive skepticism, 221

passivity, 101, 117
Paulson, Henry, 113
peer behavior, 95–100
peer groups. *See* social networks
People's Climate March, 93
perception, 206
personal behavior. *See* individual
    responsibility
personal interactions, 100–102, 125. *See
    also* social networks
personal signals of progress, 152
personal transportation. *See* transportation
personified risk, 44, 45, 58, 90
persuasive technology, 129
pessimism, ix, 218–19, 221
Pett, Joel, 141
*Plague, The*, 225–26
plane tickets, offsets for, 64, 125, 131
Plato, 209–10
*poiesis* of air, 204–6
polar vortex, 169
political action, xviii–xix, 118–19. *See also*
    voters
    individual action and, 91–92, 125
    policy change, pressure for, 104, 130
    quality of life as priority, 140–41
political parties, 68, 92
politicians
    climate-action role, 21, 96, 110, 112
    climate-change denial by, 11–13
    optimism of, 217–18
    signals for, 152
pollen counts, 114–15
pollution, 199. *See also* greenhouse gas
    emissions
    communities affected by, 174
    health problems caused by, 115, 174
    Kuznets curve, effect of, 165–66
    natural sinks for, 136
Polman, Paul, 133
preparedness climate message, 116–19
price
    carbon (*See* carbon price/tax)
    high price of climate action, 110

printer default settings, 125
produced capital, 157–58
profitability
    of climate action, 135, 138
    emissions and value added, 154
psyche
    air as, 205–10
    human, 206, 210
    of land, 207
psychic numbing, 175
psychological climate paradox,
    xviii–xxi, 3–8
psychological denial. See individual
    (psychological) denial
psychology, 181, 205–6. See also cognitive
    psychology; social psychology
    barriers and solutions, 81–84, 87–94
    conventional psychology, limits of,
        199–201
    ecological, 193, 200
    evolutionary, 27–34, 41, 100
    of identity, 70–80, 82
    principles of climate psychology, 89–92
psychotherapy, 76–78, 88, 171, 181–83,
    189–90, 192–93, 200
public support, developing, 125, 130.
    See also framing (messaging) climate
    change; political action; simple solutions;
    social networks
public transportation, 125

rainfall, changes in, 194
Randall, Rosemary, 104
Randers, Jorgen, ix–x, xvi–xviii
re-wilding story, 145–48
realism, 219
recycling, 97, 127, 153
regional initiatives, 93
regulations, 20–22
    framing of, 110, 111
    resistance to, 5, 21, 71, 73, 96, 130
religions
    beliefs and stories, air in, 211–12
    good-versus-bad dualism, 179

relationship to nature of, 142–44
renewable energy. See clean energy
resistance, ix–x, 130, 149, 186
    climate-change denial as, 76–78, 109
    to cutting emissions (See greenhouse
        gas emissions)
    to higher costs, 96, 130
    to regulations (See regulations)
resource efficiency, 136–38. See also
    energy efficiency
respiratory disease, 114
RinkWatch, 102–3
risk, insurance against, 112–14
risk perception, climate change, 42–46
    communicating, 39, 42, 46
    control and, 44, 45
    cultural identity and, 72, 73
    distance and, 39
    linking risk to opportunity, 83
    personified risk, 44, 45
risk vividness, 27, 33–34, 100
Risky Business, 113
Road, The, 28
Rome Statute, 144–45
Rothschild, Michael, 44–45
rush-hour tariff, 130

sacrifices. See losses, costs, and sacrifice,
    framing climate change as
Schellenberger, Michael, 119
Schönau, Germany, 126–27
science. See also climate science
    citizen-based, 102–3
    of ecology, 199
    limits of, 199–200
second-loop learning, 193–94
self-defense, 81–82, 90
self-destructiveness, 25, 28, 76, 91
self-governance, 102
self-interest, 27–29, 100
Sell the Sizzle, 139
shaming, 24, 60, 92, 97
Shapley, Harlow, 204
Shimkus, John, 12

short-termism, 27–28, 32–33, 100, 150,
    219–20
signals of progress, 93, 151–62
    complex information, visualizing, 152
    green-growth signals, 153–56
    Nature Index, 160–61
    personal, 152
    response to, 162
    social progress, indicators for
        measuring, 152–53
    societal transformation story, tied to, 152
    stories, connection to, 153, 162
simple solutions, 124–31
    as communication strategy, 93, 129–31
    food-waste-reduction nudges, 127–28
    green defaults, 124–27
    green nudges for homes, 126–27
    smart homes, 128–29
sinks, natural, 136
Sinnette, Kevin, 12
skepticism
    climate science, 3, 9–11, 14, 16, 31
    as variety of hope, 221–22
small-actions approach, 88–89
Snyder, Gary, 219
social capital, 157
social citizens, acting as, 91–92
social dissonance, 65, 67–69, 82
social distance, of climate change, 40
social learning, 103
social media, 98–100
social networks, x, 93, 95–109. See also
    social norms
    anti-climate-action groups, 108–9
    citizen groups, climate action by, 102–6
    communications for, 107
    eco-networks, pro-environmental
        behavior by, 105
    peer behavior, 95–100
    personal interactions, 100–102, 125
    power of, 108–9
social norms, xx, 95, 107–8, 129, 150
    climate science, scientists' consensus
        on, 101

definition, 100
    peer behavior, role of, 96–100
    signals, response to, 162
    status quo, support for, 22
social psychology, 54–69. See also attitudes,
    climate change
    Hawthorne effect, 54–55
    IPCC reports, communication style of,
        55–57
    social dissonance, 67–69
societal transformation, 19–20
solar power, 103, 111, 120, 136–38
solastalgia, 174
space, distance of climate change in, 40
States of Denial, 16
status, 27–28, 30, 100
status quo, support for, 22, 108
Steiner, John, 180
Stern, Nicholas, 113
stewardship story, 142–45
Steyer, Tom, 113
Stiglitz, Joseph, 151
Stockholm, Sweden, 130
story-based communications, x, 93, 101,
    132–50, 193. See also apocalyptic stories
    air and wind in creation story, 211
    ancient myths, 168–70
    green-growth narrative, 135–38
    inevitability of stories, 148–50
    inspiring stories, need for, 134
    methods for better storytelling, 148–50
    re-wilding story, 145–48
    signals and, 153, 162
    stewardship story, 142–45
    well-being narrative, 139–42
Strava app, 98
subsidies, 111, 138
Sullivan, John, 12
supportive messaging. See framing
    (messaging) climate change
sustainability, ix, 98, 102, 133, 143, 157. See
    also climate action
Sweden, 130, 155, 202–3
Swim, Janet, 39

Switzerland, 155
symptoms, climate disruption, 190–201
    nature's voice, listening to, 193–95
    science and psychology, limits of,
        199–201
    solution-focused approach, 191–92
    values reflected by, 198–99
    worldview and, 195–98
system boundaries, 194
systems theory, 192

tea party movement, 120
Tesla Motors, 99, 133
time, distance of climate change in, 40
Totnes, United Kingdom, 104
Transition Town Movement, 104–5
transportation, 99, 125, 130, 137
trees, 115
2050, halving emissions by, 154
*2052: A Global Forecast for the Next Forty
    Years,* ix, xvii
Typhoon Haiyan, 80

uncertainty, framing climate change as,
    110, 116–19
Unilever, 133
United Kingdom, 4, 104, 126
United States
    climate change, concern about, 4–6, 29,
        41–42, 58–59, 65–66
    climate change as politically divisive
        subject, 72
    green communities, 104–5
    politicians, climate-change denial by,
        12–13
    tea party movement, 120
United States Congress, 12–13
United States Department of Defense, 112
urgency, climate change, 22, 70, 71
    communicating, 35, 38, 42, 90, 179
    cultural cognition and, 73
    distance and, 38–39, 41, 46
    preparedness and, 118
    sacrifice, climate change framed as, 52

scientists' consensus on, 101
    social attitudes and view of, 58–59
utility service green energy programs,
    126–28

value added, 153–57
values. *See also* ethics
    attitudes and, 57, 66–67
    of audience, 56
    climate-action policy and, 118–19, 121
    climate symptoms and, 198–99
    confirmation bias and, 17, 73–75, 82
    deeper values, determining, 194
    love, value of, 187
    social networks, role of, 102, 106, 108
van der Post, Laurens, 207
Vestforbraending energy company, 128
visualization, 42–43, 60, 148, 152–53, 161
voters, 117
    carbon tax, position on, xix, 21
    climate action by, xx, 92, 108

waste
    eliminating, 136–38
    nudges to reduce, 127
water use, 97, 100, 108, 127, 137
Watts Up With That, 13, 15, 49–50
weather events, xiv, xviii, 40–42, 80,
    112–14, 167–69, 197
well-being
    counting, 156–58
    narrative, 139–42
Whitfield, Ed, 12
Williams, Raymond, 133
Wilson, Edward O., 188
wind, 206, 211–13, 225
wind power, 111, 137–38
wishful thinking, xix–xx
wolves, urban setting for, 146–47
word of mouth, 101
World Health Organization, 115
world soul, 209–10
worldview
    climate-change denial and, 19, 74

climate symptoms and, 195–200
cultural cognition of facts and,
    72–74
Wysham, Daphne, 186

Xcel Energy, 103

Zapffe, Peter, 178
Zindler, Ethan, 120

Per Espen Stoknes is a psychologist and an economist. An entrepreneur, he has co-founded clean-energy companies, and he spearheads the BI Norwegian Business School's executive program on green growth. He has previously worked both as a clinical and organizational psychologist and as an adviser in scenario planning for a wide range of major national and international businesses, government agencies, and nonprofit institutions. His research interests include climate and environmental strategies, economic psychology, and energy systems, and his teaching areas include climate strategy, foresight and corporate strategy, team development, and behavioral economics. He has written three books, including *Money & Soul*. He lives in Oslo, Norway.

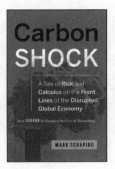
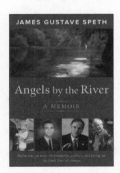

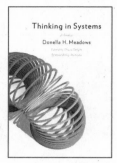